EXT-

AIRCRAFT

HarperCollins books may be purchased for educational, business, or sales promotional use. For information, please write: Special Markets Department, HarperCollins Publishers, 10 East 53rd Street, New York, NY 10022.

Produced for HarperCollins by:

HYDRA PUBLISHING
129 MAIN STREET
IRVINGTON, NY 10533
WWW.HYLASPUBLISHING.COM

FIRST EDITION

Library of Congress Cataloging-in-Publication Data

Miller, Ron, 1947–
  Extreme aircraft / Ron Miller. -- 1st ed.
     p. cm.
  Includes index.
  ISBN 978-0-06-089141-1
  1. Research aircraft--History. 2. Aeronautics--Records. I. Title.

TL567.R47M55 2007
629.133--dc22

2007023014

07 08 09 10   QW   10 9 8 7 6 5 4 3 2 1

EXTREME AIRCRAFT

Collins

An Imprint of HarperCollins Publishers

Ron Miller

Contents

The World in the Air

The extreme wonder of flight has always inspired awe in humans. The story of human flight is one of visionaries and adventurers, courageous men and women who pushed the technology to fantastic levels of speed, height, and distance. From the Wright brothers to rocket scientists, humans have embraced the ability to fly, breaking barriers with physical endurance, strength of character, and scientific proficiency.

Illustration showing a lunar halo and luminescent cross observed during the flight of the balloon *Zénith*, 1875.

Evolution of Flight

Few inventions have altered the course of human history as the airplane has—from global events to the daily life of individuals, aviation has opened the world in much the same way that the automobile freed people from living in cities. Like the wishes granted to the fisherman who liberated the genie from a bottle, the granting of the age-long wish to fly has been both a boon and a burden. It has enabled humankind access to the entire planet, but has also been a terrible instrument of war.

Aircraft not only transport us in safety and relative comfort to any point on the globe—no place on Earth is more than 24 hours away from any other place by air—they also transport mail, food, raw materials, and manufactured goods. And when time is of the essence, transport by air becomes indispensable. Cropdusters help us grow more food more efficiently, while aerial surveyors discover new deposits of oil, minerals, and metals. Doctors can quickly bring medical aid to remote regions of the world—doctors who travel by air are often the only ones that some people in remote parts of Alaska and Australia ever see. Oceanographers and meteorologists keep a close eye on the climate and weather from the air, often enabling them to save thousands of lives and billions of dollars in damages. Aircraft have even taken us beyond this planet to the edge of space.

Through most of human history, people have been unable to rise any farther above the ground than they could jump. Transportation depended on one's own feet, animals' backs, and

Opposite: Airship USS *Akron*, 1930s

eventually on animal-drawn vehicles. If a river, mountain, or canyon lay on the way to a destination, an alternate route had to be found or a bridge or tunnel constructed. Even after the locomotive and automobile allowed us to travel faster and in more comfort, we still faced many of these same limitations. Ships could traverse the seas carrying more passengers and cargo than any land-based form of transportation, but they were limited to the oceans, rivers, and harbors.

Balloons to Spacecraft

The flight of the first balloon in 1783 gave people a new perspective on the world around them. When Orville and Wilbur Wright flew the first airplane, it caused a similar seachange. Charles Lindbergh's solo flight across the Atlantic in 1927 was only one of the outstanding achievements in an era filled with awe-inspiring feats, as aviator after aviator broke records almost as quickly as they were set.

While at first the airplane seemed to be little more than a fragile novelty, Americans soon witnessed aircraft helping their allies defeat Germany in World War II. Pilots returning from the war barnstormed across the United States, thrilling and inspiring citizens in every corner of the country. It was

Opposite: X-38 under the wing of a B-52 mother ship.

the beginning of the great golden age of aviation—the whole world went aviation-crazy.

The creation of airmail service led directly to the establishment of the first U.S. airlines and, consequently, bigger, faster, safer aircraft. Technical developments during World War II gave aviation an enormous boost, with postwar aircraft manufacturers providing machines of unprecedented size, power, and luxury. The growth of the airlines changed the way people traveled and did business, and was the first step toward a global economy. There are those who now believe that manned spaceflight is essential to the survival of our species, and space exploration is seen as a truly international endeavor.

Extreme Aircraft is about inventors, engineers, pilots, the great aircraft they built and flew, and how those marvelous flying machines forever changed the world.

Chronology

1783
The first untethered balloon flight is made on November 21.

1794
Observation balloons used for the first time at the Battle of Fleurus.

1804
Sir George Cayley begins glider experiments and successfully tests a model.

1852
Henri Giffard flies 17 miles (27 km) in a steam-powered dirigible.

1853
Sir George Cayley builds a triplane glider that travels a short distance, the first recorded flight by an adult in a winged aircraft.

1861
The American Army Balloon Corps is formed under the command of Thaddeus Lowe.

1870
The French use balloons to transport letters and passengers out of besieged Paris during the Franco-Prussian War.

1872
Paul Haenlein tests the first airship equipped with a gas engine, achieving 12 miles per hour (19 km/h).

1883
Gaston Tissandier, who fits a Siemens AG electric motor to his dirigible, makes the first electric-powered flight.

1884
The first fully controlled dirigible flight is made by Charles Renard and Arthur Krebs in the electric-powered *La France*.

1891
Otto Lilienthal builds his first glider.

1897
Clement Adler claims to have made a short powered flight in his *Avion* III.

1903
In October, Samuel Pierpont Langley's *Aerodrome* fails its test flight and crashes into the Potomac River.

December 17: At Kill Devil Hills in Kitty Hawk, North Carolina, Orville Wright travels 120 feet (36.5 m) in the Wright Flyer. The flight lasts 12 seconds. Three more flights are made that day. On the best of those flights, Wilbur travels 852 feet in 59 seconds.

1905
October 14: The flying organization Fédération Aéronautique Internationale (FAI) is founded.

1906
Alberto Santos-Dumont's *14-bis* plane wins the 3,000-franc Archdeacon Prize for a flight of nearly 197 feet (60 m).

The first international balloon race takes place in Paris.

1907
Glenn Curtiss forms his aircraft company, the first in the United States.

1908
Henry Farman wins the 50,000-franc Deutch-Archdeacon Prize for the first official circular flight in Europe of one kilometer.

Charles W. Furnas becomes the first airplane passenger when he goes aloft in a Wright plane.

1909
Louis Blériot is the first to fly across the English Channel in an airplane.

Count Ferdinand von Zeppelin forms Delag (Die Deutsche Luftschiffahrt Aktiengesellschaft), the world's first commercial airline company.

1910
Elise Deroche becomes the first licensed female pilot.

Henri Fabre's *Hydravion* is the first seaplane to take off from water.

The Zeppelin LZ7 *Deutschland* begins passenger service.

First air freight is flown by Phil Parmalee on October 25.

1911
Pierre Prier makes the first non-stop, passenger-carrying flight from London to Paris.

Calbraith P. Rodgers makes the first coast-to-coast flight across the United States.

The first official U.S. airmail flight is made by Earle L. Ovington.

1912
Glenn Curtiss flies the first true flying boat, a converted Curtiss A2. Harriet Quimby becomes the first woman to cross the English Channel in an airplane.

1913
The world's first aerial combat takes place in Mexico when two

American mercenary pilots, Dean Ivan Lamb, flying for Pancho Villa, and Philip Rader, flying for President Huerta, shoot at each other.

1914

The first flight of the Sikorsky *Ilya Muromets*.

1915

French pilot Roland Garros downs a German Albatros by shooting through the steel-plated propeller of his plane.

German Zeppelin *LZ38* makes the first bombing raid on London.

1916

Pilot Manfred von Richthofen (the Red Baron) makes his first combat victory.

The Boeing Airplane Company is founded by William E. Boeing.

1918

The U.S. Army Signal Corps establishes the first airmail service in America, between New York and Washington.

Four Curtiss JN-4 Jennies complete a coast-to-coast flight across the United States, from San Diego to Jacksonville.

1919

German airline Deutsche Luft-Reederei begins the first sustained daily passenger airline service.

Three U.S. Navy flying boats attempt to make the first air crossing of the Atlantic. Only one of the planes succeeds in reaching England.

1920

United States Army Air Service (USAAS) is created. Captain Corliss C. Moseley wins the first Pulitzer Trophy Race.

1921

Experimentation with pressurized cabins begins at Wright Field.

1922

March 20: The U.S. Navy commissions its first aircraft carrier, the USS *Langley*.

1923

Lt. J. A. Macready and Lt. O. G. Kelly of the U.S. Army Air Service make the first nonstop flight across the United States.

The first demonstration of in-air flight refueling

1924

Douglas World Cruisers *Chicago* and *New Orleans* complete the first round-the-world airplane flight, arriving in Seattle after a flight lasting 371 hours and 11 minutes.

1925

U.S. Navy airship *Shenandoah* breaks in two during a storm over Ohio.

1926

Lufthansa airline is established.

U.S. Navy Lt. Cdr. Richard E. Byrd and Floyd Bennett make the first flight over the North Pole.

The first airship flight over the North Pole is made by Roald Amundsen.

1927

Pan American World Airways (Pan Am) is established.

Charles Lindbergh makes the first solo transatlantic flight.

1928

In Germany, a specially adapted glider becomes the first rocket-powered aircraft to fly, piloted by Friedrich Stamer.

1929

The *Graf Zeppelin* makes the first circumnavigation of the world by an airship.

1930

The first coast-to-coast air service in the United States is inaugurated by Transcontinental Western Air (TWA).

1931

Wiley Post and Harold Gatty fly around the world in the *Winnie Mae*.

1932

Amelia Earhart is the first woman to make a solo flight across the North Atlantic.

1933

Wiley Post makes the first solo flight around the world.

1935

The Luftwaffe is created in Germany.

1936

First flight of the Focke-Wulf Fw61, the world's first completely successful helicopter.

1937

The *Hindenburg* is destroyed by fire at Lakehurst, New Jersey.

1938

Howard Hughes completes a record round-the-world flight, taking only 3 days, 19 hours, and 17 minutes.

The first flight of the Boeing 307 Stratoliner, the first airliner with a pressurized passenger cabin.

1939

The first flight of a piloted, purpose-designed rocket-powered aircraft, the Heinkel He-176, is made in Germany.

The first flight of a turbojet aircraft, the Heinkel He-178, is made in Germany.

1941

The United States Army Air Forces (USAAF) is established.

Pearl Harbor is attacked by carrier-based Japanese aircraft.

1942

October 1: Robert M. Stanley flies first U.S. jet plane, Bell *XP-59* Airacomet.

1943

Specially modified RAF Avro Lancasters of 617 Squadron make the Dam Busters raids on the Möhne, Eder, and Sorpe dams in Germany.

1944

The first jet aircraft combat takes place when a German Messerschmitt Me-262 engages an RAF Mosquito.

1945

Japanese rocket-powered Ohka suicide aircraft score their first major successes against the USS *West Virginia* and three other ships.

1946

Pan Am inaugurates its first scheduled New York–to–London service.

1947

Capt. Charles "Chuck" Yeager breaks the sound barrier in the Bell *X-1* rocket-powered aircraft.

1948

On the 45th anniversary of the first powered flight, the original Wright *Flyer* is returned to the Smithsonian Institution by the London Science Museum.

1949

The first nonstop flight around the world is made by Capt. James Gallagher, flying the U.S. Air Force B-50 *Lucky Lady II*.

1951

An English Electric Canberra becomes the first jet to make a non-stop crossing of the Atlantic.

Alaska Air becomes the first airline to fly over the North Pole.

1953

May 18: Jackie Cochran becomes the first woman to break the sound barrier.

1954

The Convair *XFY* goes from vertical to horizontal flight and back.

1956

Milburn Apt sets a new airspeed record in the Bell *X-2*, becoming the first person to exceed Mach 3, before losing control of the aircraft and fatally crashing.

A U.S. Navy R4D Skytrain is the first aircraft to land at the South Pole.

1957

Three Boeing B-52 Stratofortresses make the first nonstop flight around the world by turbojet-powered aircraft in 45 hours and 19 minutes.

1960

USAF Capt. Joseph Kittinger sets an unofficial world record for highest parachute jump (102,200 ft/ 31,150 m) and longest parachute free fall (84,700 ft/25,815 m).

1961

April 12: Soviet cosmonaut Yuri Gagarin makes the first human spaceflight in *Vostok 1*, orbiting the Earth once in 108 minutes.

1962

John Glenn becomes the first U.S. astronaut to orbit the Earth in *Mercury* 6.

1963

The Boeing 727 makes its first flight.

1964

Geraldine "Jerrie" Mock arrives in Columbus, Ohio, in her Cessna 180, becoming the first woman to make a solo round-the-world flight.

First flight of the SR-71 Blackbird.

1965

The first B-52 Stratofortress missions fly against North Vietnam.

A Boeing 707 makes the first polar circumnavigation of the world.

1966

First flight of the Northrop *M2-F2* lifting body.

1969

Apollo 11 astronaut Neil Armstrong becomes the first man to walk on the Moon.

1971

First crossing of the Atlantic by the Concorde.

1972

Cessna builds its 100,000th aircraft, the first company in the world to achieve this figure.

First flight of the Airbus A300.

1974

An SR-71 Blackbird crosses the Atlantic Ocean in less than two hours.

1975

Svetlana Savitskaya sets a new women's airspeed record of 1,667 miles per hour (2,683 km/h) in the Mikoyan *Ye-133,* a modified MiG-25PU two-seat trainer.

1976

A Pan Am Boeing 747SP makes a record-breaking flight around the world in 1 day, 22 hours.

1977

First glide test of the Space Shuttle *Enterprise.*

The *Gossamer Condor* becomes the first human-powered airplane.

1978

Ben Abruzzo and crew make the first transatlantic crossing by balloon in the *Double Eagle II.*

The first flight of a solar-powered aircraft, the *Solar One.*

1979

Bryan Allen flies the *Gossamer Albatross* across the English Channel using pedal power.

1980

Janice Brown pilots the MacCready *Gossamer Penguin* on its first solar-powered flight.

1981

Columbia is the first operational Space Shuttle to take off.

Ben Abruzzo and crew make the first crossing of the Pacific Ocean by balloon, in the *Double Eagle V.*

1982

American Airlines flies its millionth passenger.

1984

Joseph Kittinger makes the first solo transatlantic balloon flight, from Carbon, Maine, to Savona, Italy.

1986

The Space Shuttle *Challenger* explodes shortly after launch.

Jeana Yeager and Dick Rutan make the first nonstop flight around the world in *Voyager.*

1988

Kanellos Kanellopoulos recreates the mythical flight of Daedalus by flying a pedal-powered aircraft, the MIT *Daedalus,* from Crete to Santorini.

1991

Eastern Air Lines is dissolved after 64 years of continuous operation.

Pan Am is dissolved after 63 years of continuous operation.

1993

McDonnell Douglas produces its 10,000th aircraft.

1996

The merger of McDonnell Douglas and Boeing is announced.

1999

March 1–19: Bertrand Piccard and Brian Jones make the first nonstop, round-the-world balloon flight in the *Breitling Orbiter 3.*

2001

The unmanned aircraft *Global Hawk* flies from Edwards Air Force Base to Australia nonstop and without refueling.

2003

The Space Shuttle *Columbia* disintegrates on reentry, killing all seven occupants.

The Concorde makes its last scheduled commercial flight.

2004

SpaceShipOne is the first privately built spacecraft to transport a person into space and return safely to Earth.

NASA's *X-43* reaches a record speed of Mach 10 (7,000 mph/11,200 km/h).

2005

The world's largest passenger plane, the Airbus A380, is unveiled.

Steve Fossett completes the first nonstop, solo circumnavigation of the world in the Virgin Atlantic *GlobalFlyer.*

Records

The Biggest Airplane

The Antonov An-225 Cossack is the largest powered aircraft currently in service. Built to transport the Buran orbiter, it was an enlargement of the successful An-124 Ruslan Mriya. With a maximum weight of 1.3 million pounds (600,000 kg), the An-225 is also the world's heaviest aircraft. Although its wingspan is less than that of the Hughes H-4 Spruce Goose, the latter never went beyond a single short low-altitude test flight, making the An-225 the largest aircraft in the world to take off more than once.

The Biggest Helicopter

The Russian Mi-26 helicopter, the heaviest and most powerful helicopter in the world, was designed for carrying cargoes weighing up to 20 tons. It is also used for construction projects. The Mi-26 was the first helicopter equipped with a main rotor consisting of eight blades, This was powered by two turboshaft engines. The helicopter is capable of flying on a single engine in case of a failure in one of the engines.

Ranking	Aircraft	Maximum Takeoff Weight
	The World's Biggest Planes	
1	An-225 Mriya	1,300,000 lbs (600,000 kg)
2	Airbus A380	1,230,000 lbs (560,000 kg)
3	Boeing 747-400ER	910,000 lbs (412,769 kg)
4	An-124 Ruslan	892,870 lbs (404,999 kg)
5	C-5 Galaxy	840,000 lbs (381,018 kg) (wartime max load, peacetime load is limited to 769,000 lbs/348,812 kg)
6	Airbus A340-600	837,755 lbs (379,999 kg)

The remaining engine's output increases automatically to allow continued flight.

The Smallest Airplanes

The first plane to claim the title of "world's smallest" was the *Wee Bee,* designed, built, and flown during the late 1940s by Ken Coward, William Chana, and Karl Montijo of San Diego, California. The only *Wee Bee* ever built made its first flight in 1948. Since then, aircraft have been built in an attempt to outdo the *Wee Bee*. Ray Stits and Martin Youngs built the *Stits Junior* from a surplus World War II Taylorcraft L-2 that same year. Wilbur Staib followed with the *Little Bit*. Beginning in the 1950s, Stits, and eventually his son, Donald, competed with

Robert Starr to build the world's smallest plane. The record is currently held by Starr's *Bumble Bee II*, at 8.8 feet long and with a wingspan of only 5.5 feet.

The Fastest

Few people think of the Space Shuttle as an aircraft, yet it acts as one when it reenters the

A Russian Antonov An-225 Cossack transporting a Buran orbiter

The World's Smallest Planes

Ranking	Builder	Plane	Length	Wingspan	Empty Weight	Engine	Max Speed
1	Starr	*Bumble Bee II*	8 ft 8 in (2.6 m)	5 ft 5 in (1.6 m)	396 lbs (180 kg)	85 hp	190 mph (306 km/h)
2	Stits	*Baby Bird*	11 ft (3.3 m)	6 ft 3 in (1.9 m)	252 lbs (114 kg)	55 hp	110 mph (177 km/h)
3	Starr	*Bumble Bee*	9 ft 4 in (2.8 m)	6 ft 6 in (2 m)	547 lbs (248 kg)	85 hp	180 mph (290 km/h)
4	Stits	*Sky Baby*	9 ft 10 in (2.9 m)	7 ft 2 in (2.1 m)	452 lbs (205 kg)	65 hp	185 mph (298 km/h)
5	Staib	*Little Bit*	11 ft (3.4 m)	7 ft 6 in (2.3 m)	390 lbs (132 kg)	85 hp	Never successfully flown
6	Stits	*Junior*	10 ft 10 in–11 ft 4 in (3.3–3.5 m)	8 ft 10 in–9 ft 4 in (2.7–2.8 m)	400 lbs (181 kg)	75 hp	150 mph (241 km/h)
7	Bee Aviation	*Wee Bee*	14 ft 2 in (4.3 m)	18 ft (5.5 m)	837 lbs (380 kg)	30 hp	82 mph (132 km/h)

atmosphere and maneuvers for a landing. On reentry, the Shuttle can reach speeds of up to 17,500 miles per hour (28,164 km/h). The rocket-powered X-15 was the fastest powered manned aircraft, reaching a top speed of 4,510 miles per hour (7,258 km/h) on October 3, 1967. However, like the Space Shuttle, it was incapable of taking off under its own power. The SR-71 Blackbird, with a sustained top speed of 2,188 mph (3,521 km/h), is usually considered the fastest conventional aircraft, since it was a piloted aircraft powered by conventional air-breathing engines and was capable of taking off and landing unassisted on con-ventional runways. Although officially retired in the 1960s, the SR-71 still holds aviation's top speed records.

The fastest propeller-driven aircraft is the Russian Tu-95/142, with a maximum level speed of 575 miles per hour (925 km/h), which is Mach 0.82. The Tu-95 first flew in 1954 and was initially operated by the Soviet strategic air force. The plane is powered by four turboprop engines, each driving eight-bladed contra-rotating propellers.

The unpiloted Boeing X-43A Hyper-X is the fastest air-breathing aircraft, setting a speed record of 7,000 miles per hour (11,265 km/h)—Mach 9.68—on November 16, 2004.

The Hyper-X is unmanned, relies on a carrier aircraft to lift it to the altitude it needs for launching and, like the Space Shuttle, requires a booster rocket to reach the operating speed required by its scramjet engine. It is also incapable of landing. After flying on the scramjet for approximately ten seconds, the Hyper-X goes into a 10-minute glide, after which it makes an intentional crash landing in water.

Fastest Microlight Flight
Colin Bodill's 175-hour, 4-minute flight between London and Sydney, Australia—a distance of 13,653 miles

The World's Fastest Aircraft			
Space Shuttle	manned	17,500 mph (28,164 km/h)	booster-assisted takeoff
X-43A Hyper-X	unmanned	7,000 mph (11,265 km/h)	booster-assisted
X-15	manned	4,510 (7,258 km/h)	booster-assisted takeoff
SR-71 Blackbird	manned	2,188 mph (3,521 km/h)	unassisted takeoff, jet engine
Tu-95/142	manned	575 mph (925 km/h)	propeller-driven

(21,972 km)—is the fastest journey ever made in an ultralight aircraft. Between May 31 and September 6, 2000, Bodill made an around-the-world flight in just 99 days, setting another speed record.

Fastest Balloon Flight

Steve Fossett flew around the world solo from June 19 to July 2, 2002—13 days, 8 hours, and 33 minutes—in the *Spirit of Freedom*, a Roziere-type balloon measuring 180 feet (55 m) tall. His epic flight, the fastest nonstop balloon flight around the world, started at Northam, Western Australia, and ended at Eromanga, Queensland, Australia. It was the second nonstop balloon circumnavigation, and the first flown solo.

Greatest Distance

As with all other aviation records, determining the longest flight depends a great deal on what kind of aircraft is being discussed. There is a record distance flown by every type of aircraft, from jumbo jets to paper airplanes.

On November 11, 2005, a Boeing 777-200LR Worldliner set a new world record for distance traveled nonstop by a commercial airplane when it landed at London Heathrow Airport. The 777-200LR flew 11,664 nautical miles during its 22-hour, 42-minute eastbound flight from Hong Kong, a distance exceeding halfway around the world.

The distance record for a manned balloon of 25,361 miles (40,814 km) was set by Switzerland's Bertrand Piccard and Britain's Brian Jones, when they piloted the *Breitling*

Orbiter 3 around the world from Switzerland to Egypt on March 1–21, 1999.

Highest Flight

There are many different ways to rise above the Earth—floating to great altitudes in a balloon, flying into the stratosphere in a plane, or shooting into space under rocket power. The current record for greatest altitude ever reached in an aircraft is held by a vehicle in the latter category—*SpaceShipOne*, which reached 281,310 feet (85,743 m) on June 21, 2004.

The unmanned gas balloon *Winzen*, with a volume of 1,765,733 cubic yards (1.35 million cubic meters), set the altitude record for this type of

Greatest Distance Flown		
Type of aircraft	Make	Distance
Commercial	Boeing 777-200LR	11,664 miles (18,771 km)
Balloon	*Breitling Orbiter 3*	25,361 miles (40,815 km)
Paper airplane	Ken Blackburn	27.6 seconds

Highest Fliers				
Rank by Altitude	Type	Craft/Make	Pilot	Height
1	Rocket-launched airplane	*SpaceShipOne*	Michael W. Melville	281,310 ft (85,743 m)
2	Unmanned gas balloon	*Winzen*	N/A	170,000 ft (51,800 m)
3	Manned gas balloon	*Strato-Lab* 5	Malcolm D. Ross and Victor E. Prather	113,740 ft (34,668 m)
4	Nonrocket powered	*Helios* (solar-powered)	N/A	96,500 ft (29,413 m)
5	Hot-air balloon	*Envelope*	Vijaypat Singhania	69,852 ft (21,290 m)
6	Manned airliner	NASA ER-2	Jim Barrilleaux	68,700 ft (20,940 m)
7	Piston-driven propeller airplane	*Caproni* 161	Mario Pezzi	56,047 ft (17,083 m)
8	Sailplane glider	*Grob-102*	Robert Harris	46,267 ft (14,102 m)

aircraft (and second place for all types) in October 1972. A manned balloon, the *Strato-Lab 5*, follows at 113,740 feet (183,046 m), a record set in 1961. *Helios,* an unpiloted, solar-powered airplane, holds the record for the greatest altitude achieved by a non-rocket-powered aircraft. This flying wing with a 247-foot (398 m) span covered with solar-power cells reached an altitude of 96,500 feet (29,413 m) on August 13, 2001. On November 26, 2005, Indian businessman, philanthropist, adventurer, and aviation enthu-siast Vijaypat Singhania set the world altitude record for the highest hot-air balloon flight, reaching 69,852 feet (21,290 m). While this flight set the record,

it came up short of Singhania's goal of 70,000 feet (21,336 m). A world altitude record for medium-weight aircraft was set by NASA's ER-2 aircraft on November 19, 1998. It reached 68,700 feet, nearly twice the cruising altitude of most airliners. The highest altitude obtained in a piston-driven propeller aircraft was achieved by Mario Pezzi above Montecelio, Italy, when he attained an alti-tude of 56,047 feet (17,083 m) on October 22, 1938. Sailplanes are gliders designed to rise upward on air

currents. Pilot Robert Harris set a goal to break the world glider altitude record. To prepare, he spent five years flying to ever-increasing altitudes. In February 1986, after struggling with ice, cold, and a failing oxygen sup-ply, Harris reached an altitude of 49,009 feet (78,872 m), finally returning to Earth using his backup oxygen.

Mi-26 helicopter

Strange Aircraft

Aviation has always attracted some of the most imaginative designers, engineers, and inventors—and, as is often the case, imagination surged far ahead of technology. This was Jack Northrop's dilemma when in the 1950s he developed the flying wing, a fabulous invention that required the technology and materials of the 1990s. Abandoned as impractical by the armed forces in the 1950s, Northrop's flying wings provided the foundation for the modern stealth aircraft.

Other inventors introduced extraordinary technology and engineering that resulted in unconventional design sometimes considered just too weird to be accepted by pilots, the aviation industry, or the public. And while certain aircraft were successful in spite of their unorthodox appearance, many others were produced by hopeful aviation pioneers who, unfortunately, had not the slightest idea what they were doing.

Here are some examples of aircraft that have been considered outrageous.

Lockheed P-38 Lightning

One of the most recognizable fighter planes in World War II, the Lockheed P-38 Lightning was one of the few twin-engine fighters at that time that was agile enough to survive in sustained combat. The Lightning's high speed, long range, and heavy firepower proved to be especially lethal against the Japanese Pacific forces. The plane's unusual twin-engine, twin-boom configuration has

Lockheed XFV-1

made the Lightning a favorite among the world's modelers and airplane buffs.

McDonnell XF-85 Goblin

The intercontinental range of the six-engine B-36 Peacemaker required the development of a long-range fighter that could escort it to its targets in World War II. None of the first generation of jet fighters were capable of accompanying the bomber for any distance. McDonnell Aircraft Company solved the problem with a miniature jet fighter that could be carried in the bomb bay of a B-36. Only 15 feet (4.5 m) long, the XF-85 Goblin was the smallest jet fighter ever built. Its pilot literally straddled the plane's 3,000-pound-thrust engine, which provided a theoretical top speed of 650 miles per hour (1,050 km/h). From its first test flight, in

McDonnell XF-85 Goblin

1948, it became apparent that the design concept was pushing the limits of practicality, and the XF-85 project was abandoned late the following year.

Lockheed XFV-1

The concept behind the Lockheed XFV-1 was a high-performance fighter that could take off and land vertically, like a helicopter, eliminating the need for specially prepared runways.

Both Lockheed and Convair were awarded contracts to develop the XFV-1 in 1951. The Convair version was quickly dubbed the Pogostick—a name which the Lockheed VTOL soon shared.

After a vertical takeoff, the aircraft would ease into horizontal flight, whereupon the plane could be maneuvered like a conventional fighter. At landing, it returned to the vertical position. Although the planes worked in principal, their development was plagued by endless engineering, materials, and technical problems. The Lockheed project was canceled in 1955 after making 32 flights, none of which involved vertical takeoffs or landings.

Ryan X-13 Vertijet

The Ryan Aeronautical Company developed the X-13 Vertijet for the U.S. Air Force in 1954, with the Navy and NACA joining the project later on.

The resulting plane measured only 24 feet long (7.3 m) with a wingspan of just 21 feet (6.4 m). To keep its weight to a mere 7,200 pounds (3,265 kg), the experimental aircraft had no landing gear, flaps, armament, dive brakes, or much of anything else other than the most necessary test instrumentation. The first gimbaled nozzle ever mounted in an aircraft controlled pitch and yaw so the X-13 was able to maneuver while hovering. Wingtip thrusters provided roll control.

At takeoff, the X-13 would be raised on the bed of the truck that carried it and then simply

Ryan X-13 Vertijet

lift off vertically, rise to a few hundred feet, and arch over into horizontal flight.

Bell X-5

Bell Aircraft Corporation received a contract for two variable-sweep-wing experimental aircraft in 1949. Two X-5 aircraft were constructed. These small planes could change the sweep angle of their wings in flight. The wings of the X-5 could be moved through their full range, from 20 to 60 degrees, in 20 seconds.

Swing wings allowed an aircraft to sweep its wings back for speed advantage in flight and extend them nearly straight for easy takeoff and safe landings. The X-5 made its first flight in

1951. The Air Force's F-111 and B-1 bombers, the Navy's F-14 fighter, and many aircraft types from other nations eventually took advantage of the engineering that developed from the remarkably successful X-5 program.

VZ-PAV Avro-Car

Avro Canada developed a disk-shaped VTOL aircraft in 1953. The U.S. Air Force and Army took over the project to develop an aircraft that would ascend vertically and reach flight speeds of 1,500 miles per hour (2,400 km/h). After development costs of nearly $10 million, the aircraft never rose more than 4 or 5 feet above the ground before becoming unstable.

The Avro-Car was between 18 and 25 feet (7.6 m) in diameter and

Avro-Car

weighed 3,600 pounds (1,630 kg). It was powered by three centrally mounted gas turbine engines driving a central fan used for vertical takeoff. Once it was in the air, the exhaust of the turbojet shifted to the rear to provide forward thrust.

The Aerocar

Attempts to make a combination automobile and aircraft go back to at least 1921, but the first successful flying car was invented by Moulton B. Taylor in the late 1940s. His first prototype flew in 1949 and received FAA certification in 1956, after which Taylor started manufacturing the Aerocar in quantity. Four models were produced, which accumulated more than 200,000 miles (321,890 km) of road travel and more than 5,000 hours in flight.

Bell X-5

Conversion from plane to auto could be accomplished by a single person in less than five minutes. The detached wings and tail boom were towed behind the car on retractable wheels. Unfortunately, although a successful invention, the Aerocar fell victim in the 1970s to new federal safety and emission standards, which would have made the aircraft impractically heavy.

The Guppies

The Aero Spacelines Guppy-201 is surely one of the most startling-looking aircraft ever built. Originally developed in 1961 specifically to transport large rocket stages for NASA, the plane was specially adapted from an existing Boeing B-377 Stratocruiser. The Stratocruiser's fuselage was extended 16 feet, 8 inches (5.08 m) with a bulging superstructure added that gives the plane a distinctive resemblance to a fish. Cargo as large as 9 feet, 9 inches in diameter could be loaded onto the plane. The even larger Super Guppy soon followed. Also designed to carry rocket stages, it had a greater wingspan and more powerful engines.

Four Guppy-201s were acquired by Airbus in the early 1970s for the transportation of large aircraft parts. The cargo bays of these planes are 111 feet, 6 inches (33.99 m) long, 25 feet, 6 inches (7.77 m) high, and 25 feet, 1 inch (7.65 m) wide. For loading, the entire forward section of the plane— flight deck and all—can be swung 111 degrees to one side.

Goodyear XAO-3 Inflatoplane

Perhaps the strangest of all aircraft was the inflatable rubber airplane developed in 1956 by Goodyear Aircraft. The 60-horsepower airplane could fly at a maximum speed of 70 miles per hour (113 km/h) for five hours without refueling. Designed for the U.S. Army and Navy, the plane had a one-piece wing, tail assembly, and cockpit that could be carried in a jeep or small truck when deflated, or parachuted from an aircraft to downed pilots. The entire aircraft, engine and all, weighed only 550 pounds (250 kg) but could carry 240 pounds (109 kg). Inflated by a bottle of pressurized gas in less than five minutes, the 20-foot-long (6 m), 28-foot (8.5 m) wingspan Inflatoplane could even be fitted with a hydro-ski for takeoffs and landings on water. Both single-seat and two-seat models were tested. Twelve Inflatoplanes were built during the course of the project. Development, testing, and evaluation of the inflatable airplane continued through 1972 until the project was canceled in 1973.

A Super Guppy

How Airplanes Fly

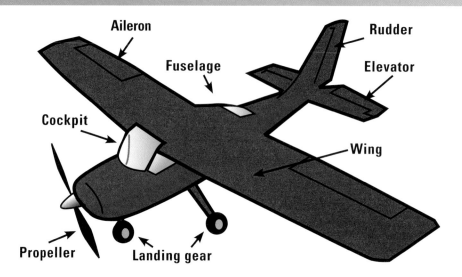

Aileron

Fuselage

Rudder

Elevator

Cockpit

Wing

Propeller

Landing gear

In order for an airplane to fly, three problems need to be solved: lift, an upward force that is greater than the weight of the plane; thrust, the force that propels the plane forward; and control, the ability to direct and stabilize the plane while it is in flight. The Wright brothers had to solve all of these. Before them, inventors had relied on sheer brute power to send their machines into the air. It was the Wrights' genius to see that humans would have to fly their machines. In Wilbur's words, "It is possible to fly without motors, but not without knowledge and skill."

Lift

Air passing over the arched upper surface of a wing must travel farther than the air passing beneath the wing. It therefore has to move faster, making the air pressure drop relative to the pressure under the wing. This creates upward lift. The degree of the wing's curvature also affects lift. The greater the angle, the greater the lift—up to a point. Beyond a certain angle the smooth flow of the air over the wing suddenly becomes turbulent and the wing "stalls" as lift is lost.

Control

A wing is inherently unstable back to front. Because lift is greatest behind the center of gravity, the wing will rotate around that point. The nose of the aircraft will then pitch down as the tail comes up. To counteract this, a horizontal stabilizer is added to the tail. This acts like an inverted wing, creating negative lift that holds the tail down.

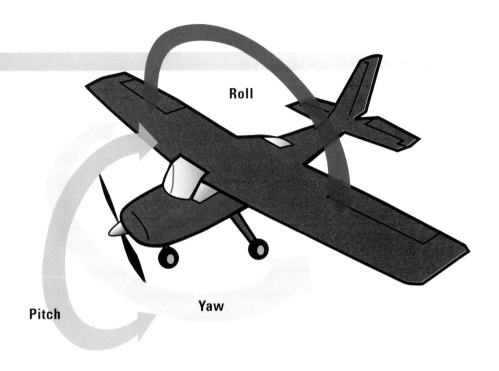

Roll

Pitch

Yaw

Lateral stability of the plane is affected by the amount of dihedral: the amount the wings are angled up or down. Movable control surfaces produce the three movements needed for maintaining control of the aircraft and changing direction: roll, pitch, and yaw. The elevator produces pitch (the up and down movement of the nose). Ailerons produce roll (the rotation of the wings around the long axis of the plane) for lateral control. The rudder controls yaw (right and left movement). These movements in combination steer the aircraft.

Thrust

Just as air flowing over the wings generates lift, airflow over the rapidly turning blades of a propeller-driven plane produces thrust, or forward motion. Each blade of the propeller acts like a wing. As the blade rotates, air flows over its curved surface. The resulting horizontal "lift" propels the aircraft forward. Because the velocity of the blade increases from hub to tip, the blade is twisted, providing the most efficient angle of attack at each point along its length.

Cross section of aircraft wing or a rotor blade.

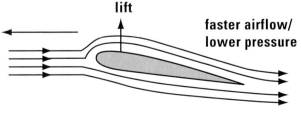

lift

faster airflow/ lower pressure

slower airflow/higher pressure

PIONEERS OF AVIATION

PART 1.

AERODROME "A"
by
SAMUEL PIERPONT LANGLEY
1903

PLAN VIEW

FRONT VIEW

SIDE VIEW

CHAPTER 1

THE FIRST INTO THE AIR

FROM THE MOMENT HUMANS first looked up at the skies, they envied the birds. Soon imaginative and daring humans had begun thinking about how they might be able to soar through the air themselves.

With no way to do this naturally, the only recourse of humans was their imaginations. Stories were told, and legends and myths grew of both humans and gods who had achieved the ability to soar above the Earth. Far from discouraging people, tales of such beings only made the goal of flight seem that much more desirable. Every human attempt to emulate a bird, however, ended in failure—sometimes disastrously so. This was in large part due to the fact that these would-be conquerors of the air badly underestimated the complexities of flight. The ability of birds to fly is an extraordinarily sophisticated achievement. It involves much more than simply beating against the air as one would make a boat move by beating against the water.

The first successful human flights, even those in gliders, resulted from the labor of many geniuses who realized that a bird's wing was more than a mere paddle and that many physical forces were involved—forces that had to be discovered, measured, experimented with, and respected.

Left: The 1904 flying machine designed by Horatio F. Phillips. Inset: A design drawing of Samuel P. Langley's Aerodrome A. Pages 18–19: The Douglas A-26B attack bomber, developed by the Douglas Aircraft Company for the Army Air Forces in 1941. During World War II, the Douglas factory produced an average of one plane every hour.

The Dreamers

If humans themselves seemed to be denied the ability to fly, they did not hesitate to assign wings to their gods and goddesses and to write innumerable stories about humans equipped with wings—both natural and artificial. As far back as c. 300 BCE, the sculptor of *Victory of Samothrace* gave this work a spectacular pair of wings. Mercury wore winged sandals that allowed him the great speed demanded by his job as messenger of the gods.

In 1638, long after the Greeks dreamed of flying, Francis Godwin (1562–1633), the bishop of Hereford, wrote a novel called *The Man in the Moone, or a Discourse of a Voyage Thither by Domingo Gonsales, the Speedy Messenger*. The hero, a Spanish sailor named Gonsales, is shipwrecked but devises an imaginative scheme to escape

his predicament: He trains a bevy of wild swans to come at his call and to carry light burdens. Eventually, he is able to attach the birds to a harness. By clinging to a kind of trapeze beneath them, Gonsales hopes he will be able to be carried to safety. Unfortunately, he is unaware that his swans belong to a very special species that every year migrates to the Moon.

Godwin's novel was not the least among the scores of published stories and schemes involving flight. Along with other fanciful fictions, there were plans for real flying machines. Most of the latter remained confined to paper, which was fortunate. Those actually tested usually suffered the fate of the flying device invented around 1678 by Besnier, a French locksmith. After failing in his own attempts to get aloft,

ICARUS AND DAEDALUS

Daedalus was the legendary architect who supposedly built the famed Labyrinth for King Minos of Crete. Myth has it that he eventually lost the favor of the king and was imprisoned in a tower. To escape, he fashioned wings of feathers for himself and his son, Icarus. The wings were held together with wax. As they launched themselves, Daedalus urged his son to stay near him, but Icarus, filled with the wonder of being able to fly, soared too near the Sun. The heat melted the wax, his wings disintegrated, and he fell into the sea, as depicted in this seventeenth-century etching.

"I hold it farre more honour to have been the First Flying Man, than to bee another Neptune that first adventured to sayle upon the Sea."

—Francis Godwin

A

D

C

B

E F

Fig. 1.

fly using muscle power alone. Da Vinci's solution was to multiply that power through a system of levers, resulting in a surprisingly modern-looking machine that lacked only an engine and propeller to resemble aircraft that flew at the beginning of the twentieth century. Unfortunately, even Leonardo's ingenious device was incapable of flight, and 600 more years would pass before a human being could finally fly by muscle power alone.

A French locksmith named Besnier designed a flying machine in 1678. It was to have been powered by the muscles of his legs and arms. He claimed to have launched himself in this device from the roof of his house and to have flown over a nearby barn, landing on the roof of another house. He sold the apparatus to a traveling showman, who was killed trying to use it.

Besnier decided to sell his contraption while he still had all his bones intact. The traveling showman to whom he sold his wings, however, was not as lucky and was killed trying to use them.

Da Vinci

When the great artist and scientist Leonardo da Vinci (1452–1519) turned his attention to the challenge of human flight, it was perhaps the first time the question had been considered by someone knowledgeable in science and engineering. He had already filled many sketchbooks with his studies of bird flight, designed one of the first parachutes in history, and come up with an idea for a helicopter. He realized that human beings could not

Among Leonardo da Vinci's many designs for flying machines is this helicopter. It was to have been powered by the action of the pilots turning the airscrew by hand.

Early Attempts

In the thousands of years of history preceding the end of the eighteenth century, no human being had ever left the Earth farther than he or she could jump, or be thrown. That is, until a pair of brothers living in France watched a column of smoke rising above a fire and started to wonder about it.

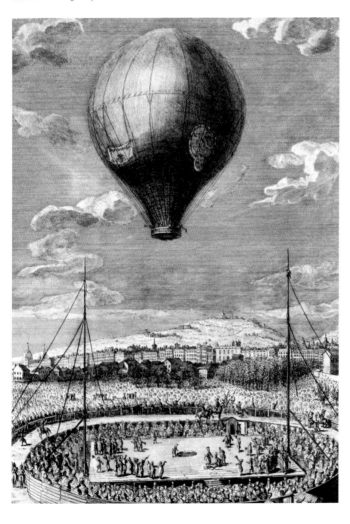

The Montgolfiers

The brothers were Joseph Michel Montgolfier (1740–1810) and Jacques-Étienne Montgolfier (1745–99), the sons of a wealthy paper manufacturer in the town of Annonay, France. The brothers, especially Joseph, devoted much time to the study of physical science, but their understanding of the science seemed rudimentary at best. Joseph had determined that air heated to 180 degrees Fahrenheit (82° C) occupied twice as much space as an equal volume of air at room temperature. He reasoned that, just as a boat floats because the water it displaces weighs more than the boat itself, a bag that displaces a volume of air that weighed more than the bag might float in the air. There was only one way to find out.

The Montgolfiers' first balloon was a small bag made of thin taffeta fabric that contained less than 78 cubic inches (1,278 cubic meters) of heated air. In November 1782 this apparatus rose to the ceiling of Joseph's apartment in what was probably the first balloon flight in history. "Prepare a supply of taffeta and cordage," he wrote to his brother, "and you shall see the most astonishing thing in the world!"

Together, the brothers tried the same experiment outdoors, this time with a much larger balloon containing more than 65 cubic feet (1.8 cubic meters) of

"Prepare a supply of taffeta and cordage, and you shall see the most astonishing thing in the world!"

—JOSEPH MICHEL MONTGOLFIER

hot air. It rose so vigorously that it tore loose from the cords holding it down and soared to a height of 200 or 300 feet (60–90 m).

Emboldened by this success, the brothers decided that it was time to make a public appearance with their new invention. For this event, they created what was essentially a huge sack, about 35 feet (10 m) tall, made of paper-lined cloth that they had fastened together with buttons and buttonholes. Burning straw filled the machine with hot smoke—in spite of what they knew about hot air, the Montgolfiers remained convinced that smoke itself was an important ingredient—and as soon as it was inflated, the contraption leaped from the ground. It soared to about 6,000 feet (1,800 m) in only 10 minutes.

Although the Montgolfier brothers were anxious to send humans aloft, they had no idea what the effect of flight on the human body might be. Therefore, on September 19, 1783, they launched a balloon with a sheep, a duck, and a rooster in its basket. The demonstration took place before an audience at the royal palace in Versailles, with King Louis XVI of France and Queen Marie Antoinette in attendance. After an approximately eight-minute flight that covered two miles (3.2 km) and obtained an altitude of about 1,500 feet (450 m), the balloon crashed to earth. The animals survived the trip unharmed, and so the brothers felt they were ready for the first human flights.

On November 21, 1783, the brothers' newest construction was viewed by half a million people, who watched in awe as the vast hot-air balloon slowly rose from the garden of the Château de la Mouette in Paris. Aboard were scientist Jean-François Pilâtre de Rozier and the Marquis d'Arlandes. The balloon rose quietly to 300 feet; a south wind then carried them five miles in 20 minutes to a safe landing in a field.

It was the first-ever flight of a (human) passenger-carrying aircraft.

The achievement of the Montgolfier brothers was hailed with the same universal adulation accorded to Charles Lindbergh 140 years later. Typical is this print showing the brothers being elevated to the status of demigods while surrounded by angels, Neptune, and Mars. A man, possibly Jacques-Étienne Montgolfier, sits atop a balloon while holding a portrait of his brother, Joseph.

The Balloon Improved

The first flight of Jacques Charles's balloon, as it ascends over the throngs crowding the Tuileries Gardens in Paris. Charles and his assistant were on this first flight, which landed two hours later, 27 miles from the city. Charles made the second flight alone. It ascended rapidly to 9,000 feet (2,750 m)—an "alarming experience," according to the inventor. It was the last balloon flight he ever made.

When news had reached Paris of the first Montgolfier flight in Annonay, Jacques A. C. Charles (1746–1823), a young scientific lecturer and experimenter, was commissioned to repeat the experiment. It did not even occur to Charles that the brothers might have filled their balloon with hot air. Being familiar with the latest discoveries in chemistry, he assumed that they had used hydrogen gas (which had been discovered just six years earlier by the English chemist Henry Cavendish). Charles knew that this gas was lighter than an equivalent mass of air. This meant that hydrogen would rise in exactly the same way as hot air. Charles realized that if it were possible to fill a large, light bag with enough of the new gas, it would rise.

Balloons over Paris

Charles began his experiments by constructing a

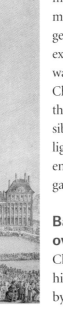

balloon made of rubber-covered taffeta about 12 feet (3.6 m) in diameter. At five o'clock on the morning of August 26, 1783, to the sound of booming cannon, the first hydrogen-filled balloon made its ascent from Paris. The vast crowd of fifty thousand people that had gathered to watch saw the balloon soar to nearly 3,000 feet (900 m). It came to earth 15 miles (24 km) away.

On December 1, 1783, Professor Charles and his assistant Marie-Noel Robert built a hydrogen balloon, which they launched from the grounds of the Tuileries Gardens. They landed two hours later, 27 miles (43 km) from Paris. Then Charles reascended, this time by himself. After rising swiftly to 9,000 feet (2,750 m), the badly shaken scientist swore he would never make another flight.

Charles's balloon introduced all the features of the modern gas balloon in one fell swoop. It not only used gas, but had valves and ballast to control the altitude of the balloon, a net to suspend the car, and so on. Although hot-air balloons continued to be built, the gas balloon was superior in every way. It could stay up a lot longer and carry much greater weight.

Le Géant

One of the most impressive balloons of the nineteenth century was built by famous society photographer Gaspard-Félix Tournachon (1820–1910), also known as Nadar. Ironically, Nadar created the biggest balloon of his time, called *Le Géant*, to raise money to build a heavier-than-air flying machine.

Le Géant stood 196 feet (60 m) high, with a gas bag 100 feet (30 m) in diameter. The bag consisted of two envelopes, one inside the other, along with a smaller balloon, called the compensator, beneath. Fully inflated, Nadar's monster could lift 9,000 pounds (4,080 kg).

Even more wonderful was the car itself. It was a two-story house, complete with living room, bedrooms, captain's cabin, lavatory, and photographic darkroom (Nadar was the first person in history to take photos from the air). On the roof was a flat observation platform.

Already a huge sensation when its first flight occurred without any mishap, *Le Géant* attracted thousands of sightseers; despite its popularity, however, proceeds from passenger fees were insufficient to support the hefty expenses involved in its upkeep.

How Balloons Work

A balloon is lighter than air. That is, the total weight of the balloon is less than the total weight of the air that it displaces. For instance, a balloon of 10,000 cubic feet will displace 10,000 cubic feet of air. Hydrogen gas in a balloon this size may only weigh 50 pounds, while the 10,000 cubic feet of air weighs 760 pounds. The balloon is therefore forced upward by the pressure of the displaced air, much in the same way that a bubble will rise in water or a boat floats on the surface of water. The amount of weight a balloon can carry is the difference between the weight of the entire balloon and the weight of the displaced air. In the case of our example, the balloon could lift almost 700 pounds, which would include the envelope, rigging, cargo, and anything else on board.

The lifting force of a hot-air balloon is negligible, because the difference in density between the hot air inside the balloon and the cooler air outside is slight. Hydrogen and helium, on the other hand, are much less dense than air, therefore balloons filled with either one of these gases can carry a great deal of weight.

After the successful balloon flights of the Montgolfiers and Charles, ballooning became an international rage, with hundreds of balloonists making public ascents all over Europe in much the same way barnstorming pilots did in the 1920s. This print illustrates an ascent made by Armand Petit in his hydrogen balloon, *Le Géant des Airs*—"The Giant of the Air."

Women and Daredevils

The first balloon had scarcely been invented when women joined men in soaring through the skies. On June 4, 1784, less than a year after the invention of the hot-air balloon, a famous prima donna of the Paris Opera, one Madame Elisabeth Thible, boarded the balloon *Le Gustave*. This was to be the first flight of the enormous hot-air balloon, which had been commissioned by Count de Laurencin at the city of Lyon and named in honor of Gustave III of Sweden, a monarch much interested in the new art of aviation. The balloonist M. Fleurant would accompany Madame Thible. Count de Laurencin had gallantly given up his place in the balloon to the opera star. The event drew an enormous crowd to Lyon's public park, including King Gustave himself, who bade Madame Thible a hearty *"Bon voyage!"*

Airborne Aria

At Laurencin's signal, the ropes holding down the 70-foot (21m), pear-shaped balloon were released, and it rose rapidly from the ground. Madame Thible was so thrilled she broke into song. People on the ground below could clearly hear selections from the diva's favorite operas drifting down from the clouds. The balloon leveled off at 9,000 feet (2,750 m) and remained at that altitude for the next three-quarters of an hour. It finally descended two miles (3.2 km) away, delivering its passengers back to earth safe and sound.

Madame Thible had just accomplished a feat only a scant handful of other human beings—and no other woman—had done. She had left the surface of the Earth and flown above it.

In the century following Madame Thible's flight, scores of women not only flew in balloons, but earned a very good living making exhibition balloon flights and parachute descents, though many paid for this daring with their lives. In fact, some of the most

Women were not far behind men in their enthusiasm for the new sport of ballooning. Indeed, some of the most popular exhibition balloonists were women.

famous, sought-after aeronauts of the nineteenth century were women.

Risky Business

Almost as soon as the balloon was invented, enterprising and daring individuals started making their living by touring the country and thrilling crowds by not only making hair-raising ascents but actually jumping out of balloons and floating to the ground in early versions of the parachute. Names such as Vincent Lunardi, the Garnerins (André-Jacques, Elizabeth, and Jean-Genevieve) and Jean-Pierre and Marie (usually known as Sophie) Blanchard were as famous then as Evel Knievel is today. It was not a safe occupation. The first aeronaut, Jean-François Pilâtre de Rozier, (1754–85) was also the first martyr to aviation when his attempt to cross the English Channel by balloon ended in disaster.

Crossing the Channel

On a cold January day in 1785, throngs of people hailed Frenchman Jean-Pierre Blanchard and American doctor John Jeffries as their balloon sailed through the air from England to France across the English Channel. Boats of every description followed their course over the waves.

The two-hour flight was not without its hazards. The balloon began to lose altitude as the coast of France approached, and after throwing overboard all of their ballast and every loose item in the basket, the aeronauts were forced to strip to their underclothes. Jeffries even offered to throw up his breakfast in order to lighten the balloon. Fortunately, a last-minute updraft saved them, and they made a safe landing on French soil, where they were greeted with an ovation worthy of conquering heroes. Louis XVI awarded Blanchard 12,000 francs and a monument was erected to honor the aeronauts, marking the spot near Calais where they landed.

Early parachutes were far more dangerous than balloons. Most exhibition parachutists made descents in contraptions like this: a wicker basket suspended below a silk canopy stiffened with flexible bows. André-Jacques Garnerin was the first to make a descent using a parachute without any such reinforcements.

Jean-Pierre Blanchard, who went from designing a muscle-powered flying machine to becoming one of Europe's premier balloonists, making the first balloon flights ever in many countries.

The Balloon at War

The first military use of the balloon was made in 1793 during the French Revolution. The revolutionary army operated a *corps d'aerostier* (corps of moored observation air balloon fliers), which first saw combat led by General Jean-Baptiste Jourdan, when it faced the Austrians and the Dutch in a battle fought near the village of Maubeuge. The aeronauts took a general officer aloft during the fighting at Charleroi, and at the Battle of Fleurus they were in the air for some nine hours, once again taking senior officers aloft to observe the fighting.

The first military use of the balloon in the United States occurred during the Civil War. Thaddeus Lowe was only one of several American balloonists who offered the services of their balloons to the Union Army. Among the others were John Wise, John La Mountain, and Ezra and James Allen. Lowe, an accomplished aeronaut, had already made an unsuccessful attempt at transatlantic flight. On the recommendation of Professor Joseph Henry of the Smithsonian Institution, Lowe was contacted by treasury secretary Salmon P. Chase, who invited him to visit the capital city. In a demonstration for President Lincoln on June 11, 1861, Lowe ascended in his balloon 500 feet (150 m) above the White House with a

The *Jean Bart*, a balloon flown by brothers Albert and Gaston Tissandier, is seen here descending over the Seine during the Siege of Paris. Balloons—both piloted and unpiloted—provided some of the only communication in or out of the beleaguered city.

telegraph key and operator. From there, he transmitted the following message over the wire that connected the balloon to the receiver below:

> Balloon Enterprise in the Air
> To His Excellency,
> Abraham Lincoln
> President of the United States
> Dear Sir:
> From this point of observation, we command an extent of our country nearly fifty miles in diameter. I have the pleasure of sending you this first telegram ever dispatched from an aerial station, and acknowledging indebtedness to your encouragement for the opportunity of demonstrating the availability of the science of aeronautics in the service of the country.
> I am, Your Excellency's obedient servant,
> T. S. C. Lowe

After further demonstrations, the president introduced Lowe to Winfield Scott, the commander of the Union Army. Lowe's first assignment was to report on Confederate troop movements near the suburbs of the capital. This proved to be a great success, and Lowe continued to observe and report on a great many military engagements over the course of the war, often under heavy fire. Lowe's rivals—Wise, La Mountain, and the Allens—also saw service during the war.

Called the Balloon Signal Service, the group had a specially assigned locomotive that moved their equipment from place to place. Lowe was in action from the time of the Peninsula Campaign of 1862 until the aftermath of the Battle of Chancellorsville the following year. The balloon corps was disbanded soon thereafter.

A 24-year-old Count Ferdinand von Zeppelin had been sent by the Prussian government as an official observer attached to the Union Army. While there, he observed Professor Lowe directing artillery from his balloon. Just before his return to Germany, von Zeppelin went on a tour of the western United States. He encountered John Steiner, who had once flown for Lowe, operating as a civilian aerial showman in Minneapolis. Count von Zeppelin made his first balloon ascent with Steiner.

The Siege of Paris

During the Franco-Prussian War, the Germans laid siege to the city of Paris, allowing no one in or out from September 1870 to January 1871. During this time, balloons allowed virtually the only communication with the outside world. Hastily manufactured in Paris railway

stations, more than 68 balloons escaped the besieged capital, carrying refugees, mail, and messages.

What has been called the "first aerial battle" occurred during the siege when Nadar—inventor of the great *Le Géant*—was heading into Paris when he met a German balloon at 10,000 feet (3,000 m). The German opened fire with a rifle, but Nadar managed to shoot his way out of the predicament, landing safely in the capital.

Professor Thaddeus Lowe had been a traveling balloon exhibitor until he offered his services to the Union Army as an aerial observer. Here he is seen making an ascent (controlled by his rope-handling ground crew) in his balloon *Intrepid* near Fair Oaks, Virginia.

Against the Wind

The word *dirigible* means steerable, and a dirigible balloon had been the goal of hundreds of inventors ever since the first balloon rose into the air. People wanted to go where they wanted to, not where the wind happened to take them. A balloon that could actually travel against the wind was the ideal goal. The first dirigible balloons took several forms. The main problem was not how to stay in the air, as the balloon did that automatically, but how to make it go in the desired direction. Sails were tried, on the theory that if a balloon floated on the air like a boat on water, then it should also be possible to guide a balloon with sails. Paddle wheels, propellers, and even oars were also tested, with no success. These

ideas, which perhaps looked good on paper, simply did not work in practice. This was largely because the early dirigible inventors had only two choices for power: muscles or steam. The first was simply too weak, and the latter required

Albert Tissandier (left) and his brother Gaston (right) in the gondola of their pioneering electrically propelled airship. The square boxes between them are the gangs of heavy batteries needed to generate the necessary current.

"These vehicles can serve no use until we can guide them.
I had rather now find a medicine that can cure an asthma."

—Dr. Samuel Johnson

heavy boilers and machinery. Inventors were certain that the creation of a lightweight, powerful engine would solve the dirigible balloon steering problem.

The Tissandier Brothers

The history of aviation seems to have had an attraction for brothers. The French chemist, meteorologist, aviator, and editor Gaston Tissandier (1843–99) was also an adventurer. He was one of the few brave individuals to escape by balloon during the German siege of Paris in 1870. Gaston's interest in meteorology led him to take up ballooning. His first flight was made from Calais in 1868, after which he made scores of scientific flights, gathering important information about the atmosphere. In April 1875 he achieved the previously unheard-of altitude of 28,215 feet (8,600 m). His two companions died from lack of oxygen, and although Tissandier survived, the experience left him deaf.

Gaston's brother, Albert Tissandier (1839–1906), was a talented artist who accompanied his brother on many aerial expeditions and provided the illustrations for Gaston's books and magazine articles.

In 1883 the Tissandier brothers equipped the airship *La France* with a Siemens electric motor, thus creating the first electric-powered flight. The

little airship, gaily painted in red and white stripes, was only 91 feet (28 m) long and 30 feet (9 m) in diameter. Current was supplied by 24 battery cells connected to the 1.5 horsepower motor. *La France* made two trial flights in October 1883, and another the following September. The aircraft showed that it was able to maintain its course in calm air and could be controlled easily by its rudder. It quickly became apparent, however, that in spite of the superiority of the electric motor to the steam engine for powering airships, the heavy weight of the batteries made it less than perfect. A more practical solution to the long-standing problem of airship propulsion was still needed.

Above: A portrait of Albert (left) and Gaston Tissandier.

Background: Sectional drawings of the electric airship that made an ascent on October 8, 1883.

Toward a Solution

French engineer Henri Giffard (1825–82) built and flew the first motor-powered dirigible airship in 1852. Giffard had constructed a three-horsepower steam engine, which he installed in a car suspended beneath his 144-foot (44 m), cigar-shaped balloon. Driving a three-bladed propeller at a rate of 110 rpm gave the airship a speed of six miles per hour (9.5 km/h).

On December 13, 1872, German engineer Paul Haenlein tested an airship equipped with an engine powered by lighting gas. It achieved a speed of 12 miles per hour (19 km/h), but the tests had to be abandoned due to a shortage of money. Later that year Haenlein flew a tethered dirigible with an internal combustion engine, the first use of such an engine to power an aircraft.

The airship *La France*, designed and built by the French military engineer Charles Renard, his brother Paul, and Arthur Krebs (a French army officer and pioneering automotive engineer), was first flown on August 9, 1884, with Krebs piloting the first fully controllable free-flight made by an airship. The flight covered 5 miles (8 km) in 23 minutes. It was the first full-circle flight, with a landing located back on the starting point. *La France* accomplished this on five other flights as well.

Santos-Dumont

Alberto Santos-Dumont, the son of a wealthy Brazilian coffee planter, was born on July 20, 1873. He began driving locomotives on his father's plantation at an early age and witnessed his first balloon ascent when he was 15, but his father forbade him to go aloft himself. When he was 18, Alberto went to Paris to study chemistry, physics, astronomy, and mechanics. He had long been fascinated by flight, and in 1898 he finally made his first balloon ascent.

Later, Santos-Dumont's almost constant appearance in the Parisian sky helped focus the world's attention on the progress of aviation. He maintained an entire fleet of airships at an airship station he had established at Neuilly. The Brazilian's first dirigible—*Santos-*

This blank form was passed out to passengers on the giant captive balloon Henri Giffard exhibited during the Paris Exhibition of 1878. For a small fee, thrill-seekers could make a balloon ascent—with the balloon safely tethered to a powerful steam winch on the ground, which wound them back to Earth once their time was up. They could record their observations on the provided form.

GRAND BALLON CAPTIF A VAPEUR
DE M. HENRY GIFFARD

Cour des Tuileries, le _____ 1878

lettres FEUILLE DE BORD N° ▨

OBSERVATIONS A TERRE

Baromètre.	Humidité relative.	Nombre de voyageurs. . . .
Thermomètre boule sèche. .	Peson.	Vent. . . . .
Thermomètre boule humide.	Manomètre.	État du ciel. . . . .

OBSERVATIONS PENDANT L'ASCENSION

HEURE	PRESSION	ALTITUDE	TEMPÉRATURE			HUMIDITÉ	VENT			ÉTAT DU CIEL	PESON	MANOMÈTRE
							DIRECTION	VITESSE				

OBSERVATIONS
GÉNÉRALES

"I am well convinced that 'Aerial Navigation'
will form a most prominent feature in the progress of civilization."

—Sir George Cayley, 1804

Dumont No. 1—was a cylinder made of varnished silk, with pointed ends, 82.5 feet (25 m) long and 11.5 feet (3.5 m) in diameter. Sixty feet (18 m) below this was suspended a wicker basket containing a 3.5-horsepower gasoline engine attached to a two-bladed propeller. In 1898, he ascended to 1,500 feet (450 m) and delighted the crowd with his ability to maneuver. In order to descend, he began releasing gas from the balloon. The intention was to replace the gas with air pumped into the balloon, so that the big sausage-shaped envelope would keep its shape. He discovered, however, that he could not replace the gas fast enough, and the balloon began to collapse. "All at once," he wrote, "it began to fold in the middle like a pocket-knife." The balloon began to fall. Looking toward the ground, he spotted some boys flying kites. "Catch my guide rope and run against the wind," he shouted. This caused the balloon to act like a huge kite and slowed its descent, allowing Santos-Dumont to land safely.

Between 1898 and 1905 Santos-Dumont built and flew 11 dirigibles, most of them powered by gasoline engines. In 1901 he won the 100,000-franc Deutsch Prize for being the first to fly an airship around the Eiffel Tower. He had only half an hour to make the round trip back to the start-

Alberto Santos-Dumont, the dapper Brazilian whose exploits with both airships and airplanes captured the public imagination in Europe and America.

ing line, which he did with just 30 seconds to spare. United States president Theodore Roosevelt was interested in the use of dirigibles in naval warfare and invited Santos-Dumont to the White House in 1904 to discuss the possibilities.

As impressive as Santos-Dumont's achievements in dirigible flight may have been, much of his fame in that area was more due to self-promotion than any major contribution he may have made to the development of lighter-than-air flight. While he was experimenting with small one-man airships (which American barnstormers were avidly copying), in Germany Count Ferdinand von Zeppelin was changing the face of aviation and producing the most impressive flying machines in the world.

Count von Zeppelin

Count Ferdinand von Zeppelin first became interested in ballooning and aviation in the 1860s, an interest that developed into the creation of the first large-scale, successful rigid airships—an invention that came eventually to bear his name.

In the summer of 1900, *LZ1* (*LZ* stood for "Luftschiff Zeppelin," with *Luftschiff* being the German for "airship"), the first of Count Ferdinand von Zeppelin's immense airships, lifted from the placid surface of Lake Constance. At 425 feet (130 m) long, it was by far the largest dirigible ever built. Climbing slowly to an altitude of 1,300 feet (400 m), it circled the lake for 20 minutes at 8 miles per hour (13 km/h).

Zeppelin's invention was the first of its kind. Unlike virtually every balloon and airship that had come before, the *LZ1* was supported by a rigid, internal structure of aluminum. Covered by a skin of specially treated linen and silk, it contained 17 separate gas bags, or cells, containing a total of some 350,000 cubic feet (9,900 m³) of hydrogen. Two aluminum nacelles, one at each end of the airship, each housed a 16-horsepower benzine-fueled engine running a 4-foot (1.2 m) propeller.

Although rightfully pleased by his triumph, Zeppelin immediately started to work on improvements. *LZ2* had greater power, and *LZ3* could reach speeds of up to 38 miles per hour (61 km/h). *LZ4* contained 500,000 cubic feet (14,000 m³) of hydrogen and

The giant dirigible *Graf* ("Count") *Zeppelin* at Lakehurst, New Jersey, on August 29, 1929, after making its historic around-the-world flight, the first for an aircraft of any type.

was equipped with engines developing a total of 220 horsepower. In 1908 *LZ4* succeeded in carrying 12 passengers on a 12-hour flight. A month later it made a 20-hour flight. In 1909 *LZ5* made a nonstop flight of 38 hours.

Feeling that his great invention was now safe enough for commercial use, Zeppelin, along with Dr. Hugo Eckener (chairman of the Zeppelin Company), created a scheduled air service with terminals in six German cities. Regular passenger service began in 1910, with the 20-passenger *Deutschland 1*. Over the next five years, the *Deutschland 1* and its later sister ships carried 37,000 passengers without a single injury. As impressive as this feat was, it is important to realize that Zeppelin had not been running a point-to-point passenger airline, but rather a kind of sightseeing operation.

The Blimp

Unlike the rigid airships of Count von Zeppelin, blimps have no supporting internal structure. For this reason, blimps are rarely very large. On the other hand, they are much less expensive to build and, where only a few passengers need to be carried, have proved to be extremely useful. Most people have witnessed blimps carrying advertisements or viewed sporting events broadcast from airships hovering

WHY ARE THEY CALLED BLIMPS?

No one really knows where blimps got their name. One theory is that there were once two categories of airships. Ones with internal structure, such as Zeppelins, were listed as "A: Rigid," while others were listed as "B: Limp." Another theory is that *blimp* comes from the sound a blimp makes when you thump it as you would a ripe melon.

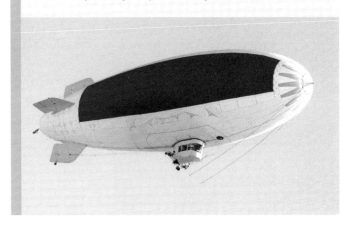

over arenas and playing fields. But blimps have had more serious uses, too. During World War I and World War II, whole fleets of slow, low-flying blimps worked as extremely effective submarine-spotters. Although today blimps are used largely for advertising purposes, geologists searching for oil and minerals take advantage of the stability of the blimp and its ability to stay aloft over a specified area for long periods. Future plans for blimps include using them to replace the expensive helicopters now used for heavy lifting in industries such as logging.

Fledgling Wings

Even though balloons gave humankind the first flights above the Earth, inventors never forgot about wings. As wonderful as the balloon was, it was at the mercy of the wind—it went where the wind went, which was not necessarily where its pilot wanted to go. Skillful balloonists could more or less steer their balloons by rising or falling until they met a current of air heading in the direction they wanted to move, but, with the exception of the dirigible, this was not a very sure way of getting anywhere.

Dirigibles, though able to be steered, were slow, expensive, and relatively fragile machines. A machine heavier than air was needed so that it would be as independent of the vagaries of air currents as birds are. In other words, what was needed was a mechanical bird of some kind.

Inventors came up with dozens of imaginative ideas, but few ever left the drawing board. The handful of early designs that were actually built never left the ground. There were two major problems facing the early creators.

Chanute 1896 biplane glider in flight at Lake Michigan.

"To fly is everything! To contrive a Flying Machine is nothing; to construct one is something; to control it in flight is to reach the heights."

—OTTO LILIENTHAL

One was finding a source of available power that was both powerful enough to move the aircraft and light enough to be carried by it. This was largely solved by the invention of the internal combustion engine. (Steam engines were relatively simple and could be tremendously powerful, but they required boilers, fireboxes, water tanks, and all sorts of plumbing in addition to the engine itself.) The second, and most difficult, problem was the development of a practical wing design and the means for controlling an airplane once it was in flight.

Pioneers

A handful of brilliant theorists and inventors laid the groundwork for the eventual triumph of the heavier-than-air flying machine. In 1799 George Cayley created the theoretical basis for the science of aeronautics and, over the next 50 years, built and successfully tested models and full-scale gliders based on his careful research. In 1848 John Stringfellow (1799–1883) created the first engine-driven model airplane that actually worked, although it had to be supported by a wire. Other inventors, such as Félix Du Temple, Alphonse Pénaud, and Victor Tatin, laid the groundwork for modern airplane design. Like others at the time, though, they lacked an adequate power source.

The invention of the four-stroke internal combustion engine by Nikolaus Otto in 1876 paved the way for the powerful, lightweight source of power that flying machines required. In 1884 Horatio Phillips (1845–1926) patented the curved airfoil, a discovery that led to the practical airplane wing. Furthermore, the hundreds of glider experiments made by German Otto Lilienthal and French-born American Octave Chanute (1832–1910) at the end of the nineteenth century were a direct inspiration for the Wright brothers.

Above: German aeronautical pioneer Otto Lilienthal in one of his biplane gliders. Lilienthal's research, conducted during hundreds of flights, provided both inspiration and foundation for the early work of the Wright brothers.

Background: Nearly a century before Lilienthal, British inventor George Cayley designed and tested model gliders. He was one of the first to approach the study of aeronautics from a scientific perspective.

Kites: The First Wings

No one is certain where kites originated. They were known to both the ancient Chinese and the early peoples of the South Sea islands, where Polynesians would fish using bait attached to a kite. Natives of the Solomon Islands still use them for fishing.

One of the earliest mentions of kites in Chinese history is of General Huan Theng, who in 200 BCE routed an enemy army by flying kites over its frightened men. Ancient Japanese and Chinese prints show fantastical images of kites carrying archers over enemy territory. Kites have also played an important role in the religion and culture of many nations.

Dr. Bell's Super-Kites

After his success with the telephone, American inventor Alexander Graham Bell began looking closely at the problem of heavier-than-air flight. In 1902 he published proof that it was possible to build a large flying machine without the need of unduly increasing the weight. Instead of building one large wing, he said, all one had to do was build many small wings arranged in the form of tetrahedrons.

Bell's greatest success with his tetrahedral kites involved one consisting of 3,393 cells. All of these cells, which weighed only about one ounce (28 g) each, were made separately from

spruce rods covered with bright red silk and measured only 10 inches (25 cm) on a side. The construction of these cells became a cottage industry for the local Nova Scotian villagers. The giant kite, named the *Cygnet*, which means swan, was towed behind a steamboat and carried a passenger, Lt. Thomas E.

Late in his life, telephone inventor Alexander Graham Bell became deeply interested in the problem of aviation. His research led him to develop his tetrahedral kites, which had enormous lifting power. Here he is seen in 1903 flying a small version. Some of his tetrahedral kites were large enough to carry a man.

Selfridge. After its first flight, landing before the steamer crew could cut the tow rope, the kite was destroyed.

How Kites Fly

The basic design of a kite is the same design used in airplane wings. Air moving over the top of the wing of an airplane creates a region of lower air pressure above the wing, because the air moving over the upper, convex surface of a wing has to travel faster than the air moving across the underside. The low-pressure area means that the region under the wing is an area of higher air pressure. The wing is drawn into the low-pressure area, creating lift. Kites fly for the same reason: air rushing over the leading edge of the kite creates a low-pressure area above the kite. The kite then tries to move into the low-pressure area, creating lift.

Although the origin of the kite is obscure, there are references to them in Chinese literature as early as 200 BCE. Kites took a long time making their way west, however, with the first known European kite dating from 1326–27. A great deal of research went into developing human-carrying kites before the invention of the airplane took over virtually all of the interest in aviation.

Ups and Downs

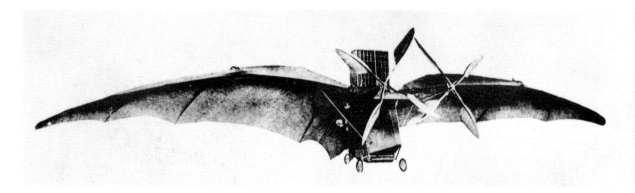

Clément Ader's infamous *Avion III*, the third flying machine built by the eccentric inventor. He later claimed to have made several successful flights in 1897 in the steam-powered machine, but most aviation historians discount this.

In his time, Clément Ader was considered an electrical and mechanical genius. In the late 1880s he turned this genius toward the problem of mechanical flight. He constructed his first flying machine, the *Éole*, in 1886. It had a batlike design and was run by a lightweight steam engine of Ader's invention, driving a four-blade propeller. The wings could be warped to control the plane in flight. Ader attempted a flight of the *Éole* on October 9, 1890, during which he succeeded in hopping along for a distance of approximately 150 feet (45 m). The plane crashed and was wrecked. While a small handful of enthusiasts consider this to be the first self-propelled flight in history, virtually all historians consider the incident to be insignificant, especially since the aircraft was clearly not under any sort of control but took off via sheer brute power. Following the wreck of the *Éole*, Ader built a second airplane. Late in his life he claimed that

he flew it in August 1892, but no proof exists that he actually did so.

Ader's *Avion*

With the backing of the French war office, Ader developed and constructed the *Avion III*. Like his other planes, it was a weird, batlike contraption. Ader attempted a flight in it on October 14, 1897. Some witnesses contended that it took off and flew a distance of more than 300 yards, while others said that it crashed before even leaving the ground. After the Wright brothers made their first flight, the French commission that had overseen Ader's experiments released reports stating that they had been successful.

Most historians discredit any claims of priority, since all of Ader's flights ended in crashes, and many of the flights were disputed. Ader himself did not help his credibility by greatly exaggerating his achievements. Ader's initial flight of the *Éole*, however, remains relatively undisputed, even though it occurred probably more by accident than by design.

Maxim's Monster

Sir Hiram Maxim (1840–1916), inventor of the machine gun, put his considerable genius and resources into the problem of heavier-than-air flight. The result was the construction of an enormous machine, 104 feet (31 m) long, weighing 8,000 pounds (3,600 kg), and consisting of several decks. The crew occupied the lowest deck, along with the steering gear, boilers, and gasoline and water tanks. Ten feet above this were the engines, each of which drove a 17-foot, 10-inch (5.3 m) propeller. Above the engines was the huge, flat main wing. To either side of the machine stretched five pairs of smaller wings. The maximum wingspan of the great airplane was 125 feet (38 m).

The huge aircraft rested on a half-mile-long (0.8 km) railway track, with a dual guardrail to keep it from rising more than 6 inches (15 cm) above the ground. On July 31, 1894, Maxim fired up his boilers and, once the giant propellers reached their full speed, gave the signal, "Let go!" The airplane raced along the tracks at 40 miles per hour (64 km/h), rising slowly into the air until stopped by the railings. Suddenly, it tore itself free and, for a few brief seconds, was airborne. Unfortunately, the timbers of the rails became entangled in the wings, and the airplane crashed to the Earth.

Apparently Maxim's sole ambition was to show that his plane could fly, but not to fly it. In addition to the restraining guardrail, the plane had running gear attached that weighed a ton and a half. He realized that no suitable method had yet been devised for controlling an aircraft in flight and that his steam engines were not capable of a sustained flight in any case. "Propulsion and lifting are solved problems," he declared. "The rest is a mere matter of time."

Sir Hiram Maxim's invention of the machine gun had provided him with the fortune necessary to indulge in his interest in heavier-than-air flight. Here he is seated at the controls of his giant flying machine.

Langley's Aerodrome

Samuel Pierpont Langley, a distinguished American astronomer and secretary of the Smithsonian Institution, was fascinated by the possibilities of heavier-than-air flight. Having developed a design based on experiments made with rubber band–powered models, he hoped to equip a full-size plane with a steam engine, but none that were light enough existed.

After successful experiments with large-scale models, he felt that he was ready to build and fly a full-size aircraft. To power this machine, Langley's chief assistant, machinist Charles M. Manly, adapted an existing gasoline engine. It weighed only 125 pounds (56 kg) and produced 52 horsepower. Langley felt he was finally ready.

Langley had been conducting his experiments from a specially constructed houseboat anchored in the Potomac River. This water site not only guaranteed him privacy, but water landings also meant that his models would not need the extra weight of landing gear. The aerodromes were launched from the upper deck by means of a large catapult.

The enormous aircraft—four times larger than his earlier models, which had 13-foot (4 m) wingspans—was ready for its first flight on October 7, 1903. Manly, clutching a life preserver and a pair of goggles, climbed into the machine. The engine was started, the two big propellers began revolving, the cable restraining the catapult was cut—and the big aircraft plummeted into the river "like a handful of mortar."

Langley rebuilt the aerodrome, tried again on December 8, but again it crashed into the river. Only nine days after Langley's second failure, the Wright brothers made the first sustained flight of a piloted, powered aircraft.

SAMUEL LANGLEY

Although Langley (pictured below) successfully flew powered, unmanned models for distances of up to half a mile in 1896, his full-size, man-carrying machine was simply too weak to fly, and Langley's approach to control of his aircraft was impractical. A number of years later, Glenn Curtiss, an archrival of the Wright brothers, successfully flew a reconstruction of the Langley machine and claimed that Langley had in fact built the first practical aircraft. In order to get it to fly any distance, though, Curtiss had to make extensive changes and improvements to the original structure.

"Few people who know of the work of Langley, Lilienthal, Pilcher, Maxim and Chanute but will be inclined to believe that long before the year 2000 AD, and very probably before 1950, a successful aeroplane will have soared and come home safe and sound."

—H. G. WELLS, 1901

The Langley *Aerodrome* on a catapult atop its houseboat carrier in the Potomac River.

SPECIFICATIONS

LANGLEY'S AERODROME
Wingspan: 48 ft 5 in (14.8 m)
Length: 52 ft 5 in (16 m)
Height: 11 ft 4 in (3.5 m)
Weight: 750 lbs including pilot (350 kg)

CHAPTER 2

THE FIRST BIRDMEN

THE FIRST HUMANS to fly were scientists who approached the problem of flight systematically. As scientists and engineers, they realized that the solution would come from a meticulous application of physics and mathematics. Ironically, the greatest of these scientist-engineers were not formally trained. The Wright brothers were amateurs—albeit perhaps among the most gifted amateur scientists in history. Instead of the trial-and-error methods employed by aeronautical researchers up to that time, the Wrights, who based their work on strong foundations laid by scientists and researchers such as Chanute and Lilienthal, started almost literally from scratch. They built up a systematic inventory of hard data and essentially invented the science of aeronautical engineering as they went along.

The Wright brothers' eventual success spurred an almost immediate explosion of enthusiasm. Aviation spread around the world, inspiring other inventors, engineers, and entrepreneurs to take up where the Wrights had left off. Technical and design innovations came almost as quickly as new airplanes could be built. Within a decade of the Wrights' first flight, airplanes were setting new altitude, speed, and distance records almost daily. The world had gone aviation-crazy.

Left: Aviator Hélène Dutrieu seated in her airplane in 1911. Dutrieu was an aviation pioneer and the fourth woman in the world licensed to fly. Inset: Henry Farman winning the Deutsch-Archdeacon Prize in 1908 for making a public circular flight of more than one kilometer and landing safely.

Lilienthal

Otto Lilienthal in flight in his 1894 glider. Photographs of Lilienthal's flights went around the world, drawing international attention to his experiments.

German civil engineer Otto Lilienthal (1848–96) built and flew gliders, making hundreds of successful flights between 1891 and 1896. He controlled his gliders by leaving his body free to move in any direction. By swinging his body and legs from side to side, Lilienthal was able to shift the center of gravity and thereby exercise a measure of flight control. Lilienthal built both monoplane and biplane gliders, and often launched himself from an artificial hill he had specially constructed. He was experimenting with flight control surfaces when he was killed in 1896. News of his death was

one of the events that led the Wright brothers to think about heavier-than-air flight.

Chanute

Octave Chanute (1832–1910) was a respected American engineer, most of whose career was devoted to railroad engineering and bridge building. This experience is reflected in the design of his box kite–like biplane gliders, the construction of which resembled bridges more than the birdlike Lilienthal gliders. After conducting experiments in his Chicago laboratory, Chanute and his two assistants, Augustus M.

SPECIFICATIONS

LILIENTHAL BIPLANE
Wingspan: 17 ft 1 in (5.2 m)
Wing area: 105 sq ft (32 sq m)
Maximum wing length: 6 ft 11 in (2.1 m)
Total glider length: 15 ft 8 in (4.7 m)

Herring (1865–1926) and William Avery, moved their operations to the windswept dunes of Lake Michigan in 1896.

Chanute was displeased with the way in which Lilienthal was forced to constantly keep his body moving in order to keep his machine balanced. Chanute believed that a flying machine needed automatic stability. That way the pilot would be able to focus his attention on steering. After two years of work on the dunes, Chanute ended up with a biplane glider that could carry a man through the air at speeds of 20 to 40 miles per hour (32–64 km/h). More than 2,000 flights were made in Chanute's gliders without a single accident.

Octave Chanute, a pioneer in the development of gliders, was not only an inspiration to the Wright brothers but also actively encouraged their work.

The Wright Brothers

Wilbur (left) and Orville Wright during their European tour.

Orville (1871–1948) and Wilbur (1867–1912) Wright worked together from the time they were boys. Both passionate about anything mechanical and excellent mechanics and machinists themselves, the brothers made a perfect team.

They had long been following the experiments of German glider pioneer Otto Lilienthal, and the news of his death in 1896 spurred their ambitions to create a flying machine of their own. As methodical as one of their own machines, the Wrights asked the Smithsonian Institution for a list of books about aeronautics, and they read everything they could get their hands on about what had been done in aviation during the preceding 40 years. They also wrote to the men whose research they most admired, such as Samuel Pierpont Langley and Octave Chanute. The latter would turn out to be their most enthusiastic mentor.

Gaining Control

As experienced engineers, the Wrights knew almost instinctively that there would be more to a successful flying machine than merely getting off the ground. It would have to be controlled if it were to be successful. They were not happy with the weight-shifting technique used by Lilienthal to try to maintain control, a method that led to his fatal accident. Inspired by watching how birds correct their flight, they came up with the idea of wing warping. This method utilized a mechanism that twisted the wingtips, changing the angle with which they met the air. In 1899 they tested wing-warping on a five-foot (1.5 m) biplane kite.

Confident that their design was sound, the Wrights built a 17-foot (5 m) glider with an elevator mounted ahead of the wings. They went to Kitty Hawk, North Carolina, hoping to gain flying experience. Unfortunately, the wings generated less lift than expected, so they flew the glider mostly as a kite, working the control surfaces from the ground. Wilbur spent a total of only 10 seconds aloft in free flight. They went home discouraged but were convinced they had achieved lateral and longitudinal control.

Trying to overcome the lift problem, they increased the camber, or curve, of the wing of their 1901 glider. They also lengthened its wingspan to 22 feet (6.7 m), making it the largest glider anyone had ever attempted to fly. When they tested it at Kitty Hawk, lift was still only a third of that predicted by the Lilienthal data upon which the wing design was based. Also the glider pitched wildly, climbing upward only to stall. When they returned to the earlier curvature, they achieved longitudinal control and eventually glided 335 feet. Despite this success the 1901 machine was still unpredictable. When the pilot raised the left wing to initiate a right turn, the machine instead tended to slip to the left. This failure, and the realization that their work had relied on false data, brought them to the point of quitting.

They decided to dump all of Lilienthal's data, which they had been using for wing design. Building and operating their own wind tunnel, the Wrights gathered much more, and much more complete, data than anyone else had to date. That data was the basis for their 1902 glider, which was a genuine success and opened the way to powered flight.

Their 1902 machine embodied all of their new research. They gave it efficient 32-foot wings (9.7 m) and added vertical tails to counteract yaw. The pilot moved a hip cradle to warp the wings. Some 400 test glides proved that the design was workable, even if still flawed. Sometimes, when the pilot tried to raise the lowered wing to come out of a turn, the machine instead slid sideways and spun into the ground. Orville suggested a movable tail to counteract this tendency. After the tail movement was linked to the warping mechanism, the

plane could easily be turned and stabilized. Unlike previous inventors who had been misled by an analogy with boats and thought that all that would be necessary for steering an aircraft was a rudder, the Wrights realized that control and stability were interrelated. After 600 more tests that year, they were satisfied that they had a working airplane.

Adding Power

Now the Wrights had to power their aircraft. Although gasoline engines had been around for decades, technology had only just recently advanced to where its use in airplanes was feasible. Unable to find a suitable lightweight commercial engine, the brothers

Wilbur Wright just after landing an experimental glider at Kitty Hawk, North Carolina, in 1901. He can be seen lying prone on the lower wing.

designed their own (with the assistance of Charles Taylor, a machinist in their bicycle shop). It was cruder and less powerful than the one used by Langley, but the brothers realized that if the wings and propellers were designed efficiently, relatively little power would be needed. It was good design versus

Above: This image shows the first successful flight of the Wright Flyer at the moment it left the ground.

Background: Front view of the 1903 Wright Flyer.

the brute force approach everyone else had used. Since such propellers did not exist, the Wrights used their air tunnel data to design the first effective airplane propeller, one of their most original and brilliant engineering achievements.

Returning to the Kill Devil Hills, they mounted the engine on the new 40-foot, 605-pound Flyer (12 m, 274 kg) that had also been outfitted with a double tail and elevators. The engine powered two chain-driven pusher propellers (one of the chains was crossed to make its propeller rotate in the opposite direction to the other in order

to counteract a tendency of the plane to twist in flight). A balky engine and broken propeller shaft caused delays, but the Wrights were finally ready on December 14, 1903. They tossed a coin to see who would be the first to fly the plane, and Wilbur won. He lost his chance, however, when he overcontrolled after leaving the launching rail. The Flyer climbed too steeply, stalled, and dove into the sand. The first flight would have to wait until after the plane was repaired.

Success

Three days later, the Wrights were ready for a second attempt. The 27-mile-per-hour (43 km/h) wind was stiffer than they would have liked, since their predicted cruising speed was only 30 to 35 miles per hour (48–56 km/h). The headwind would slow their ground speed to a crawl, but they proceeded anyway. They signaled the volunteers at the nearby lifesaving station that they were about to try again. Now it was Orville's turn.

Remembering Wilbur's experience, Orville positioned himself and carefully tested the controls. The engine was started, and at 10:35 Orville released the restraining wire. The Flyer moved down the rail as Wilbur steadied the wings. Again, the Flyer was unruly, pitching up and down as Orville overcompensated with the controls. But he kept it aloft until it hit the sand about 120 feet from the rail. Flying into a 27-mile-per-hour (43 km/h) wind gave Orville a ground speed of only 6.8 miles per hour (11 km/h), and a combined airspeed of 34 miles per hour (55 km/h). The brothers took turns flying three more times that day, getting a feel for the controls and increasing their distance with each flight. Wilbur's second flight—the fourth and last of the day—was an impressive 852 feet (260 m) in 59 seconds.

"They have done it! Damned if they ain't flew!" said an awed witness.

This was the real thing, transcending the powered hops and glides others had achieved. The Wright machine had flown. But it would not fly again; after the last flight it was caught by a gust of wind, rolled over, and damaged beyond easy repair. With their flying season over, the Wrights sent their father a matter-of-fact telegram modestly reporting the fact of their epochal achievement:

The genius of the Wright brothers lay not only in their perseverance, but in their scientific approach to the problem of flight. Superb, self-trained engineers, they developed an original research strategy that enabled them to overcome problems the existence of which previous experimenters had not even suspected. The Wrights researched what they needed to accomplish and compiled data from their meticulous experiments, thus creating a new process of technological development instead of relying on the simple method of trial and error on which their predecessors had depended. The development of the airplane evolved though logical, progressive, evolutionary stages from kites to gliders to three successful powered aircraft (1903, 1904, and 1905).

The invention of the airplane was a fundamental turning point in history. It redefined the way in which wars were fought, revolutionized travel and commerce, and fueled the process of technological change.

SUCCESS FOUR FLIGHTS THURSDAY MORNING ALL AGAINST TWENTY ONE MILE WIND STARTED FROM LEVEL WITH ENGINE POWER ALONE AVERAGE SPEED THROUGH AIR THIRTY ONE MILES LONGEST 57 SECONDS INFORM PRESS HOME CHRISTMAS
OREVELLE WRIGHT

SPECIFICATIONS

THE 1903 WRIGHT FLYER
Wingspan: 40 ft 4 in (12.29 m)
Length: 21 ft 3/8 in (7.41 m)
Height: 9 ft 3 1/4 in (2.82 m)
Empty weight: 605 lbs (274 kg)
Power: 12-hp gasoline engine

The Rivals of the Wrights

Almost immediately after the invention of the airplane by the Wright brothers, other inventors began creating aircraft, most attempting to improve upon the Wright designs. After 1903 the French built flyers based on the Wright gliders. But by 1906 none had remained aloft for more than a few seconds of uneven flight. Not until 1907 did a European plane stay in the air as long as the Wrights had in 1903. By that time the Wrights had already evolved beyond their original 1903 Flyer and, in their flights of 1904 and 1905, had proved that they could stay aloft as long as they wished. It was not until 1908 that European aviators could make the same claim.

The story of the Voisins and the Farmans continues the theme of brothers pioneering aviation. Henry Farman (1874–1958) became a French aviator and aircraft designer and manufacturer, as did his brother Maurice (1877–1964). The young Henry trained as a painter and soon became fascinated by the many new mechanical inventions appearing at the end of the nineteenth century. Farman was one of the first customers of Gabriel (1880–1973) and Charles (1882–1912) Voisin when they began selling airplanes in 1907. During that autumn, Farman made a number of modifications to his Voisin aircraft. Among other things, he reduced the size of the tail surfaces,

Gabriel (right) and Charles Voisin were among Europe's leading pioneer aviators. The brothers formed a highly successful aircraft manufacturing company and created one of the most significant aircraft of the time in 1907—the Voisin pusher biplane.

removed one of the forward elevators, and added a slight dihedral to the wings.

With this plane, the *Voisin-Farman 1*, Farman won the Deustch-Archdeacon Prize for the first recorded powered flight in a one-kilometer circle in Europe. Taking off and never flying higher than 40 feet (12 m), he flew around a pylon 1,640 feet (500 meters) away and then returned to his starting point. Farman, like Santos-Dumont, was hailed as a pioneering hero, even though the Wrights had done far better several years earlier. The Wright brothers' secrecy, engendered by a desire to protect their patents, was already starting to work against them.

Other Worthy Adversaries

Alberto Santos-Dumont (1873–1932) had by this time turned his interest toward heavier-than-air flight with the same enthusiasm he once had for ballooning. In 1906 he took the wicker nacelle (enclosure) of dirigible No. 14 and added a fuselage and box kite–like biplane wings. A 24-horsepower V8

engine drove the propeller. He dubbed the plane the *14-bis* (*bis* is French for one-half). The *14-bis* had a wingspan of 40 feet (12 m) and a 33-foot-long (10 m) fuselage, with tricycle landing gear. On August 21, 1906, Santos-Dumont made his first successful flight to an altitude of about 23 to 43 feet (7–13 m). The aircraft was badly damaged on landing. The following October, the *14-bis* flew a distance of 197 feet (60 m) at a height of 6.5 to 10 feet (2–3 m) during a seven-second flight. With this flight Santos-Dumont won the 3,000-franc Archdeacon Prize for the first flyer to achieve a level flight of at least 82 feet (30 m).

Henry Farman in his Voisin-Farman biplane, an improved version of the *Voisin 1.* On January 13, 1908, he flew the first officially observed closed circle of one kilometer in one minute, winning the 50,000-franc Deutsch-Archdeacon Prize.

Brazil's Favorite Son

As far as much of the world knew, the *14-bis* was the first airplane flight ever and Santos-Dumont became an international hero. Reports about the Wright brothers' earlier flights were scarcely believed—even in the United States—at the time. To this day, Brazilians proudly, albeit mistakenly, consider Santos-Dumont the inventor of the first practical airplane.

The last aircraft designed by Santos-Dumont was the *Demoiselle*, a tiny monoplane with a wingspan of only 16.7 feet (5 m). The entire aircraft weighed only 242.5 pounds

Pioneering aviation inventor, engineer, and entrepreneur Glenn H. Curtiss, seated in one of his biplanes.

(110 kg), including Santos-Dumont himself, and was capable of speeds over 60 miles per hour (97 km/h). Santos-Dumont refused to patent the design and instead released the drawings of the *Demoiselle* for free in the belief that aviation would eventually bring about world peace. If this was his goal, he was to be sadly disappointed. Santos-Dumont was seriously ill and despondent about the use of aircraft in warfare and committed suicide on July 23, 1932.

The Fastest Man Alive

Glenn Curtiss (1878–1930) was a champion motorcyclist, who was known as "the fastest man alive." He was first introduced to aviation when balloonist Thomas Baldwin ordered one of Curtiss's gasoline engines for his small dirigible, the *California Arrow*. In 1908 Curtiss made his first flight piloting Baldwin's airship.

Alexander Graham Bell eventually asked Curtiss to join his Aerial Experiment Association (AEA), which had been created to develop new aircraft designs. Curtiss made his first airplane flight on May 21, 1908, his 30th birthday. The plane, equipped with a Curtiss engine, was the first to be controlled by ailerons instead of the wing warping preferred by the Wright brothers. Curtiss's next plane, the *June Bug*, made flights of more than 3,000 feet and, on July 4, 1908, won the first Scientific American trophy for flying success and its $2,500 purse.

After Dr. Bell disbanded the AEA, Curtiss formed his own company. Curtiss's flights attracted huge crowds—some 300,000 people recorded at one event—as he won race after race. Curtiss's celebrity status attracted the attention of the Wright brothers, who sued him, accusing him of infringing on their patents. This endeavor wasted much of the Wrights' time and money and allowed other inventors to gain ground on them in aircraft development.

Perhaps Curtiss's greatest accomplishment was the invention of the seaplane, also known as the flying boat. His first attempt at building such a vehicle involved nothing more com-

plex than affixing a set of pontoons to his Curtiss biplane, which he dubbed the *Flying Fish*. On January 18, 1911, a Curtiss plane became the first to take off and land on a ship at sea. In February of the same year, Curtiss became the first person to take off and land on water in his new tractor seaplane.

During World War I, Curtiss developed aircraft for transportation and military purposes. He specialized in training, observation and patrol, seaplanes, and flying boats. The Curtiss Aeroplane & Motor Company, the second company that Curtiss established, would eventually become one of the world's largest aircraft companies.

SPECIFICATIONS

CURTISS D, 1910
Wingspan: 26 ft 3 in (8 m)
Length: 25 ft 6 in (7.7 m)
Engine: Curtiss 4-cyl 40-hp for training;
Curtiss 8-cyl 60/75/80-hp for exhibition

A Decade of Achievement

In the early twentieth century, crowds flocked to aviation events, such as this air show at Belmont Park, New York.

In the decade following the invention of the airplane, it seemed that everyone wanted to see what an airplane could do, and there was no shortage of courageous—and often foolhardy—men and women only too eager to fulfill the public's curiosity. Their feats may have been little more than stunts, and dangerous ones at that, but they served the very important purpose of raising public consciousness about aviation. Eventually, airplanes and everything about them became all the rage, and the men and women who flew them became the superstars of their age.

In 1909 Frenchman Louis Blériot (1872–1936) flew across the English Channel in an event that stunned both the French and the British—especially the latter, who realized that their

island nation was no longer isolated from Europe by its surrounding seas. This insight forever changed the geopolitical climate of Europe.

Even the dour Wilbur Wright was not above a publicity stunt. In 1909 he made the first airplane flight New York City had seen, flying from Governors Island around the Statue of Liberty and back, and later a round trip from Governors Island to Grant's Tomb. Since flying above any water was still an extremely risky business, Wilbur took the precaution of carrying a canoe beneath the lower wing.

Around the Eiffel Tower

That same year, Count Charles de Lambert (1865–1944), flying a Wright plane, made the first airplane flight around the Eiffel Tower. Firsts followed firsts at an incredible rate, with a record hardly set before it was outdone by some new pilot and his airplane.

A CAT AND A PRESIDENT

The first cat to cross the English Channel by air was a tabby named Miss Paris, the mascot of pilot John B. Moisant. And the ever-adventurous former president, Theodore Roosevelt, was the first such executive to fly in an airplane. Roosevelt joined pilot Arch Hoxsey in his Wright Flyer for a flight over St. Louis, Missouri, on October 11, 1910.

"It is hard enough for anyone to map out a course of action and stick to it, particularly in the face of the desires of one's friends; but it is doubly hard for an aviator to stay on the ground waiting for just the right moment to go into the air."
—GLENN CURTISS, 1909

SPECIFICATIONS

WRIGHT EX *VIN FIZ*
Wingspan: 31 ft 6.5 in (9.60 m)
Length: 21 ft 5 in (6.53 m)
Height: 7 ft 4 in (2.23 m)
Weight: 903 lbs (410 kg)
Engine: 35-hp Wright

One of the great sensations of the decade was the nonstop flight made by Glenn Curtiss from Albany, New York, to New York City in 1910. The 142.5-mile (229 km) journey was the longest flight in the Western Hemisphere. Curtiss covered the distance in 2 hours and 50 minutes at an average speed of 54.8 miles per hour (88 km/h). Along the way, he dropped down to just 300 to 400 feet (90–120 m) above the ground to race a special train that had trouble keeping up with him.

That same year saw the first non-stop round-trip flight over the English Channel by Charles Stewart Rolls (1877–1910) in a Wright biplane and a round-trip flight, by Curtiss pilot Charles K. Hamilton, from New York to Philadelphia that won a $10,000 prize.

Women pilots were still a novelty, so just about anything they did was judged an achievement: Belgian-born Hélène Dutrieu (1877–1961) of France (she married a Frenchman in 1922 and became a French citizen) set a number of records among women: She made the first hour-long flight by a woman, was the first woman to carry a passenger, and set an altitude record while carrying a passenger (1,300 feet!). Katherine Stinson and Ruth Law, the first women stunt pilots in America, toured the country giving aerobatic exhibitions.

Coast to Coast

While others were setting greater and greater city-to-city distance records, Calbraith Rodgers (1879–1912) broke them all by flying in his Wright EX *Vin Fiz* from New York City to Long

Cal Rodgers takes off from Sheepshead Bay, New York, in his Wright model EX *Vin Fiz* at the start of his 49-day coast-to-coast flight.

Beach, California, in 1911. Rodgers's incredible 4,231-mile (6,800 km) flight required 69 hops and 49 days, with only 3 days, 10 hours, and 4 minutes actually spent in the air. During the trip Rodgers crashed 19 times and did not win the $50,000 prize from William Randolph Hearst due to the fact that he did not complete the flight within one month, as the rules had stipulated.

Above the Channel

There were almost as many prizes being offered as there were pilots willing to try for them, but Louis Blériot accomplished a genuine first and what was perhaps the most important event in aviation until World War II.

The *Daily Mail* of London had offered a cash prize to the first pilot to fly across the English Channel. Blériot decided he could do this, even though his newly built, self-designed plane had given him much difficulty.

Blériot had to face two rivals for the prize. The first was Hubert Latham (1883–1912), a Frenchman whose family was of British descent. Admired

Rodgers's Wright EX *Vin Fiz* on display at Pasadena, California, November 6, 1911, after his transcontinental flight from Sheepshead Bay, near New York City.

SPECIFICATIONS

BLÉRIOT XI

Span: 28 ft 6 in (8.7 m)
Length: 25 ft 3 in (7.7 m)
Height: 8 ft 4 in (2.5 m)
Weight: 700 lbs (317 kg) loaded
Engine: Anzani 20 hp (3 cyl)

Blériot's feat demonstrated that England was, in the words of H. G. Wells, "No longer an island." Transcontinental travel was suddenly possible, and the barrier between England and Europe disappeared. British newspapers warned that airplanes flying over the Channel could carry bombs as easily as they could passengers. "Britain's impregnability has passed away. . . . Air power will become as vital as sea power," one newspaper warned.

Louis Blériot at the controls of his airplane, the model that he flew in his historic crossing of the English Channel.

by the people of both countries, he was favored to win. The other was Charles de Lambert, a Russian aristocrat and one of Wilbur Wright's best students.

In July 1909 the three competitors arrived at Calais, France. Latham attempted a crossing on the 19th, but engine trouble forced him into the sea miles from shore.

De Lambert, meanwhile, crashed while testing his plane and withdrew from the race. While Latham was preparing for another try, Blériot watched the weather. At 4:40 on the morning of July 25, he took off for England. Although he did not have a compass, Blériot successfully landed in England just 36 minutes later.

The Women Pilots

Beautiful, glamorous, an accomplished actress and skilled journalist, Harriet Quimby did much to advance public consciousness of aviation.

From the time that Elisabeth Thible went aloft in a balloon, women have taken to the skies. Blanche Stuart Scott (1889–1970) was the first woman to drive a car cross-country from New York to San Francisco in 1910. Later that same year she became the first woman to pilot an airplane. By 1930 there were hundreds of licensed female pilots.

The Glamorous Harriet Quimby

Born in Michigan, Harriet Quimby (1887–1912) was raised on a farm until her family moved to California in 1887. Beautiful and talented, she was working as an actress in San Francisco by the time she was 25 while writing for several publications. Quimby moved to New York in 1903 to work for Frank Leslie's *Illustrated Weekly*, a popular periodical for which she wrote reviews and advice articles, including articles advising women on their careers, auto repairs, and household tips. Quimby also worked as a photojournalist and traveled around the world. Her writing included screenplays for the pioneer filmmaker D.W. Griffith.

In October of 1910, Quimby attended the Belmont Park International Aviation Tournament and was immediately taken with flying. There she met Matilde Moisant and her brother, John Moisant. John and his brother Alfred

ran a flying school where Harriet and Matilde Moisant began taking flying lessons. Being a canny journalist who realized the value of publicity, Quimby made sure that the press covered her lessons. She even wrote about them herself for *Illustrated Weekly*.

Quimby won her license in 1911, becoming the second woman in the world to earn one (the first being the Baroness Raymonde de Laroche in 1910). Her friend Matilde became the second female licensed pilot in the United States. At once Quimby began touring in the United States and Mexico as an exhibition flier. In her custom-designed purple satin flying outfit, the glamorous aviatrix quickly became a sensation.

Harriet sailed to England in March, 1912. She borrowed a 50-horsepower monoplane from Louis Blériot and took off for France on April 16. Landing safely about an hour later, she became the first woman to fly across the Channel in a solo flight.

On July 1, 1912, Quimby was taking part in an exhibition flight back in the United States when the plane she was in suddenly overturned, spilling Quimby from the cockpit to fall to her death. Ironically, although she had been championing the use of seat belts by aviators in a series of articles, she was not wearing one on her fatal flight.

Against All Odds:
Bessie Coleman

Bessie Coleman (1892–1926) was born in a one-room cabin in Atlanta, Texas, in 1892. After finishing eighth grade, Coleman entered Oklahoma's Colored Agricultural and Normal University but was eventually forced to drop out due to lack of money. After working as a laundress, Coleman moved to Chicago to join her two brothers. She was working as a beautician when America entered World War I and her brothers joined the segregated Eighth Army National Guard Unit in France. When they returned, they teased their sister about the many famous women aviators they had heard about in France. Bessie became determined to earn a pilot's license herself.

Coleman's race worked against her, however, and all of the white pilots and flying schools she applied to refused to teach her. The editor of a black newspaper in Chicago encouraged her, however, suggesting that she attend a flying school in France, where such prejudices did not exist. With his help, Coleman obtained a passport, visa, and a crash course in French. She was accepted by the École d'Aviation des Frères Caudron, France's most prestigious flying school. After nine months of rigorous ground school and flight training, Coleman earned her license on June 15, 1921.

When she returned to the States, she was greeted as a novelty by the white public but as a genuine heroine by the black community. Coleman used the celebrity generated by her nationwide exhibition flights to advance racial equality in America. Unfortunately, all of the impediments of racial prejudice worked against her dream of establishing a flying school. Coleman died in 1926 when she was thrown from a plane that had gone out of control.

Bessie Coleman was the first black American woman to earn her pilot's license, though she had to go to France to learn to fly.

Hélène Dutrieu in her aviator's outfit. She had once caused a stir by flying in her corset. A serious air racer, on December 31, 1911, she won France's Coupe Femina for the women's world nonstop flight record—158 miles (254 km) in 178 minutes.

FEMALE FIRSTS
- First to fly: Elisabeth Thible, 1784
- First to solo in a balloon:
  Jeanne Labrosse, 1798
- First to solo in an airplane:
  Thérèse Peltier, 1908
- First woman to make an hour-long flight:
  Hélène Dutrieu, 1909
- First woman to receive a pilot's license:
  Baroness Raymonde de Laroche, 1910
- First American woman to pilot an
  airplane: Blanche Stuart Scott, 1910
- First American woman to receive a pilot's
  license: Harriet Quimby, 1911
- First woman airmail pilot:
  Marjorie Stinson, 1918

Flight without Fixed Wings

The helicopter as a concept is nearly as old as the airplane. Leonardo da Vinci included a drawing of one in his notebooks along with one of a flying machine, both of which were powered by human muscles. In the mid-1800s a society was formed in France with the sole goal of creating a working helicopter. Jules Verne was a member of the organization, and the novel he based on his experiences, *Robur the Conqueror*, about an inventor and his huge super-helicopter, inspired a young Russian named Igor Sikorsky (1889–1972), who swore that one day he would build a flying machine just like the one in the book. But in the years leading up to Sikorsky's realization of that dream, the helicopter had a long way to go before one actually flew.

Fixed Wing versus Rotary Wing

First, to clear away commonly held beliefs about helicopters: They are not wingless, and they do not fly by blowing a blast of air downward. A helicopter's big rotor blades are its wings. The main difference between these and the wings of a conventional aircraft is that where the entire airplane must move through the air in order for its wings to generate lift, a helicopter moves only its wings by spinning them rapidly through the air. The blades then generate lift in exactly the same way that any other wing does. Helicopters are often referred to as "rotary wing" aircraft to distinguish them from "fixed wing" aircraft.

The Autogiro

A fascinating precursor, or perhaps cousin, of the helicopter was the autogiro. Although it had a rotary wing like a helicopter, the wing was not powered. Instead, it provided lift while the autogiro was propelled by a conventional propeller. These aircraft are capable of very short takeoffs and landings, but they cannot hover like a true helicopter. The first successful autogiro was built and flown in 1923 by Juan de la Cierva.

The modern incarnation of the autogiro is the X-wing aircraft that the National Aeronautics and Space Administration (NASA) has developed. An X-wing resembles two long wings crossed at a right angle. At low speeds, the wing pivots like a helicopter rotor,

This Rotor Systems Research Aircraft (RSRA) X-wing was developed by Sikorsky Aircraft for a NASA program between 1970 and 1988. The goal was to create an efficient combination of the vertical lift of a helicopter with the high cruising speed of fixed-wing aircraft.

Vought-Sikorsky 300. The first free flight was in May of 1940. The VS-300 was not the first successful helicopter to fly, but it was the first with the single-rotor configuration that eventually became the world standard.

Inspired by the success of the VS-300, the U.S. Army awarded Sikorsky a contract to develop a military helicopter. The result, the XR-4, made its first flight in January of 1942. After a series of successful tests, the XR-4 became the world's first production helicopter.

Igor Sikorsky (left) and Orville Wright pose in front of Sikorsky's XR-4 helicopter in 1942. The XR-4 was the prototype for the world's first mass-produced, single-rotor helicopter.

SPECIFICATIONS

SIKORSKY XR-4
Rotor diameter: 38 ft (11.6 m)
Length: 33 ft 11.5 in (10.36 m)
Height: 12 ft 5 in (3.78 m)
Empty weight: 2,010 lbs (913 kg)
Engine: 175-hp Warner R-500-1 Super Scarab

generating lift and helping to propel the plane. At high speeds, however, the X-wing is locked into place, where it then acts like a conventional wing.

Igor Sikorsky

A Russian native who became a naturalized U.S. citizen in 1928, Igor Sikorsky was a pioneering aeronautical designer. His company eventually designed one of the first aircraft to attempt a New York–to–Paris flight and later built the Pan Am Clipper flying boats. Sikorsky began experimenting with helicopter-type flying machines while still in Russia. He continued to experiment after emigrating to America, and in 1938 he succeeded with the first flight of the

No Propellers

The earliest attempts at jet engines were Henri Coanda's *Coanda-1910* aircraft, Secondo Campini's *Caproni CC.2*, and the Japanese Tsu-11 engine, which was intended to power Ohka kamikaze planes at the end of World War II. None of these engines were very successful, and in some cases they were slower than conventional piston engines. In fact, Coanda's experimental planes were incapable of actual flight.

Jet Engines

The key to the development of a successful jet airplane was the gas turbine engine. This type of engine was invented in 1872, but an operating model was not built until 1903. In principle, the gas turbine engine is relatively simple—it extracts energy from the flow of hot gases. Gas burned in a combustion chamber spins a turbine, which in turn operates a compressor that compresses the air and fuel entering the combustion chamber. The rotary motion of the turbine can be used to spin a propeller, but the hot exhaust gases can also be used to create thrust. In principle this is straightforward, but in practice it requires materials that are capable of withstanding extremely high temperatures and powerful forces. Such materials and engineering did not exist at the beginning of the twentieth century.

It was not until the 1930s that engineers were finally able to start developing practical turbojet engines. On January 16, 1930, British inventor Frank Whittle (1907–96) patented his design for a turbojet engine, followed by German Hans von Ohain (1911–98) who started work on another design in 1935.

The First Jet Plane

Ohain approached aircraft manufacturer Ernst Heinkel about his design, and after a relatively short period of time, their first turbojet engine was running. A perfected version was fitted to a Heinkel He-178 aircraft and flown by Erich Warsitz on August 27, 1939. The He-178 was the world's first jet plane. Toward the end of World War II, research and development into jet aircraft technology had culminated in the Messerschmitt Me-262, the world's first operational jet fighter.

Whittle W2/700 engine. Frank Whittle developed the first turbojet engine with enough operating thrust to power an aircraft in 1939. The W2 was the second, more powerful, version of a flight-ready turbojet engine developed by Whittle. The W2/700 engine flew in the Gloster E.28/39, the first British aircraft to fly with a turbojet engine, and the Gloster Meteor.

The fighters escorting Allied bombers could not stop the Me-262, which could fly 540 mph (870 km/h) at 20,000 feet (6,100 m). Entering the war much too late to make any real difference, the Me-262 had an engine life of only 30 hours, which prevented its widespread use.

Meanwhile Whittle's Power Jets Limited started to receive financial backing from the British government. An engine called the W-1 was fitted to a Gloster E28/39 and flown on May 15, 1941. The Gloster Meteor, Britain's first jet fighter, entered the war at almost the same time as the Me-262. The Meteor did not have much impact, though, because it was 100 miles per hour (160 km/h) slower than the German jet.

The Meteor was the only operational Allied jet fighter used during World War II. The Americans had been working on jets since the late 1930s, but with little official backing or enthusiasm. This lack of interest slowed the development of early American turbojet technology. It was not until 1941 that the Bell P-59 Airacomet first flew, and work on the Lockheed F-80 Shooting Star began as late as 1943. The Shooting Star was not in operation until after the end of World War II, and it did not see active service until the Korean War, when the first jet-to-jet combat occurred between an F-80 and a Russian MiG.

SPECIFICATIONS

MESSERSCHMIDT ME-262
Crew: One
Length: 34 ft 9 in (10.58 m)
Wingspan: 41 ft 0 in (12.5 m)
Height: 12 ft 7 in (3.83 m)
Empty weight: 8,400 lbs (3,800 kg)
Engine: Two Junkers Jumo 004B-1 turbojets, 8.8 kN (1,980 lbf) each
Maximum speed: 540 mph (870 km/h)
Armament: Four 30-mm MK 108 cannons (A-2a: two cannons), two 550-lb (250 kg) bombs (A-2a only), 24 2.2-in (55 mm) R4M rockets

SPECIFICATIONS

GLOSTER METEOR
Crew: One
Length: 44 ft 7 in (13.59 m)
Wingspan: 37 ft 2 in (11.32 m)
Height: 13 ft 0 in (3.96 m)
Empty weight: 10,684 lbs (4,846 kg)
Engine: Two Rolls-Royce Derwent 8 turbojets, 3,500 lbf (15.6 kN) each
Maximum speed: 415 mph at 10,000 ft (665 km/h at 3,050 m)
Armament: Four 20-mm British Hispano cannons

CHAPTER 3

WAR IN THE AIR

THE POTENTIAL FOR THE USE OF AVIATION in warfare became apparent with the invention of the balloon. However, except for a few shots exchanged between balloonists, for more than a century aircraft were not been used for anything more than passive observation, such as when the Union Army employed Professor Lowe's balloons to report on enemy troop movements during the U.S. Civil War. Even after the invention of the practical heavier-than-air flying machine, most of the world's armies never considered using aircraft offensively. At the beginning of World War I, however, both sides were regularly using airplanes to report on enemy positions, including planes equipped with cameras. Even radios, as heavy and clumsy as the equipment was at the time, were occasionally used, which allowed pilots to direct artillery fire.

A few pilots had started carrying weapons—handguns, grenades, and even steel darts—to use against ground troops. Eventually, planes were equipped with bombs, and the creation of an entirely new type of pilot and airplane became necessary. Their specialty would be the destruction of enemy bombers and reconnaissance aircraft. Before long, bombers were supplied with escorts, and pilots were fighting one another for mastery of the skies above the front lines.

Left: The pilot of a Grumman F6F Hellcat fighter throws himself from the cockpit to escape a fire that started when an auxiliary gas tank broke loose from his aircraft, skidded across the flight deck of the USS *Ticonderoga* and was slashed by the propeller. Inset: A U.S. Army Bell UH-1 (Huey) helicopter landing in a clearing in a Vietnamese jungle.

Perfecting the Fighter

Ace German fighter pilot Manfred von Richthofen, dubbed the Red Baron by the enemy pilots who respected his flying skill, at his squadron's aerodrome after returning from a mission in 1916. Seen behind him is the German cross marking his plane.

In the early days of air combat, a fighter pilot was limited to shooting at targets on either side of his aircraft and behind. If he tried to shoot straight ahead, he risked destroying his propeller with his own bullets. French aviator Roland Garros (1888–1918) worked on a solution to this limitation. Garros had become the world's first fighter pilot when he took to the air in defense of Paris the moment war was declared in 1914. In 18 days, he brought down five enemy planes. The metal wedges Garros installed on his propeller deflected any bullets that struck the spinning blades. However, the ammunition could just as easily be deflected into the engine or, worse yet, into the pilot. In spite of this, Garros brought down an enemy fighter on the first flight of his new plane.

German engineers then began to work on the problem of shooting through the propeller without hitting it. They linked the gun to the engine with a device called a synchronizer, which prevented the gun from shooting when the propeller blade was in the way. The instrument was perfected by Anthony Fokker and applied to his aircraft. The resulting Fokker E1–EIII series became the first true fighter aircraft. Fokker pilots such as Oswald Boelcke and Max Immelmann dominated the skies over the Western Front in the spring of 1915.

The Red Baron

Baron Manfred von Richthofen (1892–1918) transferred from the cavalry to aviation at the beginning of the war. His first flights were as an observer, gunner, and bombardier. He soloed on Christmas Day of 1915. Flying an Albatros fighter, he had his first aerial duel in November of 1916. While one of his squadron's trademarks was aircraft tails painted red, the baron's red plane was known as the Red Battle Flyer, and he was henceforth called the Red Baron. A careful, methodical, professional pilot who considered anything other than bringing down enemy planes in the most efficient manner possible to be a waste of time, he eventually became the greatest ace of the war with a total of 80 victories.

The legendary Red Baron met his death on April 21, 1918. He had been engaged in a dogfight with a pair of British Sopwith Camels when a bullet fired from the ground passed through his chest. Von Richthofen managed to make a controlled landing in his Fokker triplane, but he was dead by the time Allied soldiers reached him.

SPECIFICATIONS

ALBATROS D-III
Crew: One
Length: 24 ft 0 in (7.33 m)
Wingspan: 29 ft 6 in (9.00 m)
Height: 9 ft 6 in (2.90 m)
Empty weight: 1,532 lbs (695 kg)
Engine: One Mercedes D.III (160 hp)
Maximum speed: 103 mph (165 km/h)
Armament: Two 7.92-mm LMG 08/15
  machine guns

The Sopwith Camel

The Sopwith Camel became one of the best-remembered Allied aircraft of World War I due to its agility in combat. It was notoriously difficult to handle, and pilots claimed that the airplane offered a choice between a "wooden cross, Red Cross, and Victoria Cross." The Camel and Britain's S.E.5a aircraft achieved aerial superiority over the German Albatros scouts. The Camel was credited with shooting down 1,294 enemy aircraft, more than any other Allied scout plane, with the Camel flown by Maj. William Barker becoming the most successful Royal Air Force fighter aircraft during World War I, shooting down 46 aircraft and balloons between September 1917 and September 1918.

SPECIFICATIONS

FOKKER D.VII
Wingspan: Upper: 29 ft 3.5 in (8.93 m)
Lower: 22 ft 10 in (6.86 m)
Length: 23 ft (7.01 m)
Height: 9 ft 3 in (2.82 m)
Empty weight: 1,540 lbs (700 kg)
Engine: 160-hp Mercedes or 185-hp BMW

The SPAD

The SPAD (named for the Société pour L'Aviation et ses Dérivés, the company that built it) was most notably flown in American air service by Capt. Eddie Rickenbacker and the 94th Aero Pursuit Squadron, who were known as the "Hat in the Ring Gang" for the symbol painted on their aircraft. With 26 victories, Rickenbacker was America's highest scoring ace of World War I. The SPAD was also flown by most French aces, including Georges Guynemer and René Fonck, the highest-scoring French ace.

SPECIFICATIONS

SOPWITH CAMEL
Crew: One
Length: 18 ft 9 in (5.71 m)
Wingspan: 28 ft 0 in (8.53 m)
Height: 8 ft 6 in (2.59 m)
Empty weight: 930 lbs (420 kg)
Engine: One Clerget 9B 9-cyl rotary engine, 130 hp
Maximum speed: 115 mph (185 km/h)
Armament: Two 0.303 in (7.7 mm) Vickers
    machine guns

"Aviation is fine as a sport. But as an instrument of war, it is worthless."

—Gen. Ferdinand Foch, professor of strategy,
École Superieure de Guerre, 1911

The Fokker D.VII

The Fokker D.VII was a late entry into the World War I arena. When introduced into combat in 1918, however, it quickly proved superior to existing Allied fighter aircraft. The D.VII could dive without fear of structural failure and was noted for its ability to climb at high angles of attack while resisting stalling and spinning. These characteristics contrasted sharply with Allied aircraft such as the Camel and SPAD. The D.VII was so infamous that in the armistice agreements following the war it was the single weapon specifically demanded by name to be turned over to the Allies.

Major Hartney of the 27th Aero Squadron climbing out of his SPAD XIII after returning from France.

The Bombers

The Sikorsky *Le Grand*, with its four 100-hp engines, was capable of speeds up to 60 miles per hour (96 km/h). It laid the foundation for the giant bombers developed during World War I. Its features included dual controls, electric lights, and even an open-air balcony! Opposite: The Sikorsky *Le Grand* made its first flight in May 1913 over Petrograd (St. Petersburg), Russia.

With its designer Igor I. Sikorsky at the controls, *Le Grand*, the world's first four-engine airplane, made its maiden flight on May 13, 1913, in St. Petersburg, Russia. Sikorsky had already made a name for himself designing outstanding single-engine airplanes and had realized that large, multiengine airplanes would be needed to carry passengers and freight over long distances. The four-and-half-ton *Le Grand* was truly enormous, with a wingspan of 92 feet (28 m)—three-fourths the distance covered on the Wright brothers' first flight. It was powered by four 100-horsepower engines mounted between the narrow biplane wings, two on each side.

Ahead of the pilot's compartment stood an open-air balcony. Behind the pilot was a passenger compartment with four seats, sofa, and washroom—unheard-of luxuries in a passenger airplane at the time.

Le Grand made a total of 53 successful flights, including a world endurance record of 1 hour, 54 minutes while

SPECIFICATIONS

SIKORSKY S-22
Crew: Four to eight (up to twelve)
Length: 57 ft 5 in (29.8 m)
Wingspan: Upper: 97 ft 9 in (29.8 m);
    Lower: 68 ft 11 in (21 m)
Empty weight: 6,930 lbs (3,143 kg)
Powerplant: Four Sunbeam Crusader V8
    engines, 148 hp each
Maximum speed: 68 mph (109 km/h)

carrying eight passengers. The plane was retired after that, but Sikorsky had proved his point: that a multiengine plane could be flown successfully. In doing so, he laid the groundwork for both commercial passenger airliners and future bombers.

Sikorsky's Ilya Muromets

The Sikorsky S-22 Ilya Muromets, the first mass-produced bomber aircraft in the world, was a massive four-engine biplane that could penetrate deep into the enemy territory and drop a significant load of bombs. Sikorsky based the design on his earlier S-21 Russky Vityaz, which had played an important role in the development of Russian aviation and the multiple-engine airplane industries of the world.

The Zeppelin Staaken
bomber was a genuine
giant. In a series of 32
raids on London, the
four-engine plane delivered
2,772 bombs weighing a
total of 196 tons.

First conceived of as a luxury airplane, the Ilya Muromets was the first passenger airplane to have an isolated cabin, electric heating and lighting, comfortable wicker chairs, a bedroom, lounge, and even a toilet. It was test-flown on December 10, 1913, and on June 21–23 set a world record by making a round-trip from St. Petersburg to Kiev in 14 hours and 38 minutes with just one landing.

When World War I began, Sikorsky converted the plane into the world's first purpose-designed bomber. It could carry up to 1,764 pounds (800 kg) of bombs. Positions for up to nine machine guns were also added, as well as armor for the engines.

In August of 1914 the Ilya Muromets was adopted by the Imperial Russian Army, and on December 10 of that year the Russians formed their first 10-bomber squadron. They increased the number of

squadrons to 20 by the summer of 1916. The heavy bombers developed during the war by other nations all resembled the Ilya Muromets to some degree. In fact, the design had been officially licensed to the British and French. The Russians were the first in aviation history to perform bombing from heavy bombers, group bomber raids, and night bombing. The Ilya Muromets took part in more than 400 missions and dropped a total 65 tons of bombs during the war.

The Zeppelin Bombers

The largest airplanes flown in the war, the giant German bombers were the first large planes to be designed and flown for the specific purpose of conducting long-range bombing raids.

A prewar proposal for a huge, six-engine seaplane capable of transatlantic flights came to the attention of Count von Zeppelin, who had also become

A CLOSE CALL

Only three days after the armistice was signed, the Zeppelin *LZ72* was completed. Built "for the express purpose of crossing the Atlantic Ocean and dropping four-and-a-half tons of bombs on New York City," it was surrendered to France and renamed the *Dixmunde*.

interested in developing airplanes. Realizing that a similar plane could be used as a long-range bomber, he created a company to develop just such a plane. The final result was the Staaken RV, the "Giant." It lived up to its name, too, with a 138-foot (42 m) wingspan and a length of 75 feet, 7 inches (just over 23 meters; the wingspan of the World War II Flying Fortress was 104 feet, or 31.7 meters). It was powered by five 245-hp engines, one in the nose and two each in a pair of nacelles carried between the vast wings, driving three 14-foot (4.2 m) propellers. It weighed 14 tons, of which nearly 4 tons was payload. The plane carried a crew of nine, all equipped with parachutes, which were still a rarity at that time. The Staaken defended itself with six machine guns.

The Germans used Zeppelins to successfully bomb London and other Allied cities during World War I. A total of 53 raids were made during the course of the war, with London the target in 12. The beleaguered city was also bombed 19 times by aircraft—in all, 275 tons of bombs were dropped on England during the course of the war.

British pilots made a specialty of shooting down the giant airships, which, being filled with flammable hydrogen, were extremely vulnerable to incendiary attacks. R. A. J. Warneford was the first British pilot to down a Zeppelin, dropping a bomb on it from above.

Zeppelin dirigibles made an average of two bombing raids a month on London. The material damages were not as great as the psychological effect on the civilian population.

THE ZEPPELIN RAIDS : THE VOW OF VENGEANCE
Drawn for 'The Daily Chronicle' by Frank Brangwyn ARA

'DAILY CHRONICLE' READERS ARE COVERED AGAINST THE RISKS OF BOMBARDMENT BY ZEPPELIN OR AEROPLANE

World War II

 Following World War I, the 1919 Treaty of Versailles banned Germany from having an air force. Using light training planes, the country began secretly training pilots for a future war. Initially, civil aviation schools were used in order to maintain the illusion that the pilots were training for commercial airline work.

On February 26, 1935, Adolf Hitler ordered Hermann Göring to reinstate the Luftwaffe, breaking the Treaty of Versailles. The Luftwaffe took the opportunity to test its pilots, aircraft, and tactics in the Spanish Civil War of 1936–39, when the Condor Legion was sent to Spain in support of the anti-Republican government revolt led by Francisco Franco. The aircraft included names that would

Two German Dornier 217 bombers pass over the burning targets of the Breckton Gas Works at Silvertown, a suburb in the southeast of London, in autumn 1940 during the Battle of Britain. The German bombing of England was devastating, and the country was saved only by the dogged heroism of its Royal Air Force.

become world famous: the Junkers Ju 87, the Stuka dive-bomber, and the Messerschmitt Bf 109 fighter. Germany was careful to mark these aircraft as part of Franco's Nationalist air force in order to keep the world from realizing that they had been developing a new, modern air force.

At the outset of World War II, the Luftwaffe was one of the most modern, powerful, and experienced air forces in the world, dominating the skies over Europe with aircraft that were much more advanced than those of the nations it was attacking.

Fighters in Europe and Beyond

Germany's invasion of Britain in the summer of 1940 was a turning point in the war. For the first time in history the fate of a nation rested on a battle that would be won or lost in the skies overhead instead of on a battlefield or at sea. In 1940 Germany launched its powerful air force—fresh from victories in Poland, Belgium, and France—at Great Britain. A similar

"Only air power can defeat air power. The actual elimination or even stalemating of an attacking air force can be achieved only by a superior air force."

—ALEXANDER P. DE SEVERSKY, AVIATION PIONEER

victory over the island nation seemed assured. The Luftwaffe would be pitting some eighteen hundred bombers against half that many Royal Air Force (RAF) interceptors.

The Battle of Britain was an engagement between three of the most famous fighter aircraft of World War II: the British Hawker Hurricane and Supermarine Spitfire, and the German Messerschmitt Bf 109.

Designed by R. J. Mitchell and produced by Supermarine, the Spitfire had a higher top speed than the Hurricane

SPECIFICATIONS

MESSERSCHMITT BF 109

Wingspan: 32 ft 6.5 in (9.92 m)
Length: 28 ft 8 in (9.04 m)
Height: 8 ft 6 in (2.59 m)
Empty weight: 5,900 lbs (2,700 kg)
Maximum speed: 387 mph (630 km/h)
Service ceiling: 38,500 ft (11,600 m)
Range: 425 miles (700 km)
Engine: Daimler-Benz DB 605A-1 1,475-hp, 12-cyl inverted-vee, liquid-cooled
Armament: Two 13-mm machine guns, one hub-firing 20-mm cannon, and two 20-mm cannon in underwing pods

and other contemporary designs. Much loved by its pilots, the Spitfire saw service throughout the conflict in all theaters of war, and in many different variants, with some Spitfires remaining in service well into the 1950s. It was the only fighter aircraft to be in continual production before, during, and after the war.

The Hurricane was developed in response to the British Air Ministry's call for a fighter aircraft built around a new Rolls-Royce engine. The design, started in early 1934, was the work of Sidney Camm (1893–1966). By some measures, Camm's design was outdated when introduced. It used traditional Hawker construction techniques that had been used on previous biplane aircraft. It had a fabric covering over a tubular metal frame and all-metal wings. With its simple construction, ease of maintenance, wide landing gear, and easy flying characteristics, the Hurricane was vital in times when reliability was more important than performance.

The Messerschmitt Bf 109 had been designed by Willy Messerschmitt (1898–1978) in the early 1930s. One of the first modern fighters of the era, it boasted such advanced features as all-metal construction, a closed canopy, and retractable landing gear. It was the standard Luftwaffe fighter

SPECIFICATIONS

SUPERMARINE SPITFIRE MK VB
Wingspan: 36 ft 10 in (11.23 m)
Length: 29 ft 11 in (9.11 m)
Height: 11 ft 5 in (3.63 m)
Empty weight: 5,065 lbs (2,297 kg)
Maximum speed: 374 mph (601 km/h) at 13,000 ft
    (3,962 m)
Service ceiling: 37,000 ft (11,277 m)
Range: 1,135 miles (1,826 km)
Engine: Rolls-Royce Merlin 45, 1,440 hp,
    vee 12 cyl, liquid-cooled
Armament: Two 20-mm cannon and four .303-cal.
    machine guns, external bomb load of one 500-lb
    (226 kg) or two 250-lb (113 kg)

This replica of the famed 1941 Hawker Hurricane honors one of the most celebrated fighters of World War II. It was the first monoplane in RAF service and, together with the Spitfire, was largely responsible for British victory in the Battle of Britain.

during the war, scoring more kills than any other aircraft in history. The Me-109 was flown by Erich Hartmann (1922–93), the top scoring fighter ace of all time with 352 confirmed victories. Hartmann refused to fly any other airplane in combat throughout the war. By the end of the war more Bf 109s had been produced than any other fighter aircraft in history, with over 31,000 built.

Striking first at shipping in the Channel, in the hope of drawing the British Hurricane and Spitfire fighters well away from their home bases, the Germans soon moved onto bombing the air bases in the south of England. Although the pilots of the Royal Air Force fought back with extraordinary skill and courage—the average age

being scarcely older than 23—the defenders were soon losing more pilots than they could replace. At this point the Germans could have swept the RAF off the map, clearing the way for their planned invasion, but they made a fatal mistake. Beginning on September 7, 1940, the Luftwaffe turned its attention away from harassing the RAF bases and began bombing London. Over 67 nights, an average of 163 bombers a night dropped 14,000 tons of high explosives and 12,000 incendiary bombs on the city.

With the attention of the Luftwaffe diverted from the vital British airfields, the RAF was free to defend its homeland, and over the next nine days the Germans lost 175 bombers. On October 12, Hitler, realizing that his plan to wipe the RAF from the skies was not going to succeed, abandoned his invasion of England, and the Battle of Britain was over. Prime Minister Winston Churchill

expressed the gratitude of the British people when he said that "Never in the field of human conflict was so much owed by so many to so few."

The Flying Tigers

Like the American pilots who flew in the Lafayette Escadrille before the United States entered World War I, the pilots of the Flying Tigers flew their P-40 Warhawks for China against Japanese invaders. The all-volunteer force of 100 pilots, led by General Claire Chennault (1893–1958), scored 300 victories over the enemy at a loss of only 50 planes and nine pilots on its own side. After nine months of service, the Flying Tigers were incorporated into the U.S. Army Air Force.

Left: British Prime Minister Winston Churchill was an outspoken supporter of the air defense of Great Britain, saying in 1940 that "Victory, speedy and complete, awaits the side which first employs air power as it should be employed. . . . [Germany] missed victory through air power by a hair's breadth. . . . We ourselves are now at the crossroads."

Background: The distinctive shark-faced fighters of the Flying Tigers volunteers.

From Busting Dams to Flying Wings

In 1941 British inventor-scientist Barnes Wallis (1887–1979) came come up with an extraordinary idea. He believed that if the huge dams in the Ruhr Valley were destroyed, unleashing vast quantities of water into the Ruhr industrial sector, this would play havoc with the industries supporting the German war effort. Wallis believed that an explosive charge detonated next to the wall on the lakeside of a dam would destroy it.

Since the Germans had defenses set up to foil a torpedo type of bomb attack, Wallis came up with an idea he called "childishly simple." He created a bomb that bounced over the water

SPECIFICATIONS

AVRO LANCASTER
Wingspan: 102 ft 0 in (31.09 m)
Length: 69 ft 4 in (21.13 m)
Height: 19 ft 7 in (5.97 m)
Empty weight: 36,900 lbs (16,738 kg)
Maximum speed: 287 mph (462 km/h)
Service ceiling: 24,500 ft (7,470 m)
Range: 1,660 miles (2,670 km)
Engine: Four 954.5-kW (1,280 hp) Rolls-Royce
    Merlin 24s, 12-cyl vee engines
Armament: Ten 0.303-in machine guns and up to
    14,000 lbs (6,350 kg) of bombs

British scientist, engineer, and inventor Barnes Wallis, photographed in 1943. One of Wallis's many inventions was the bouncing bomb used by the so-called Dam Busters. Wallis also developed the Wellington bomber with its innovative geodesic construction and the "earthquake bomb" used to penetrate the vast concrete German U-boat pens. After the war Wallis performed considerable work in the design of swing-wing aircraft.

like a skipped stone. It would hit the dam and then sink to the right depth before exploding.

Wallis's bomb was 50 inches in diameter, 60 inches long, and weighed 9,250 pounds. It would be dropped from a Lancaster bomber. RAF Squadron 617, the crew that came to be known as "The Dam Busters," needed to release the bomb while flying at exactly 220 mph exactly 60 feet above the surface of the water. This required exceptionally precise flying. The extreme low altitude flown by the planes during the attack was demonstrated by a Lancaster that had to turn back. It had lost its bomb when it hit the sea on the journey to the European mainland.

Nineteen Lancaster bombers took off on the night of May 16, 1943. The Möhne, Eder, and Sorpe dams were the primary targets, with the Lister and Eneppe dams as secondary. Five of the 19 planes were held in reserve during the attacks.

An hour after takeoff, the Möhne Dam was breached, followed soon by the Eder Dam. The Sorpe Dam, however, held out. Although the raid caused tremendous damage, production in the Ruhr Valley was interrupted for only a short time. The greatest effect of the raid was in drawing tens of thousands of German soldiers away from the front lines to defend the dams.

30 Seconds over Tokyo

At a time when all news from the Pacific war seemed to be bad, Lt. Col. James "Jimmy" Doolittle (1896–1993) led a spectacular raid on the Japanese homeland that raised cheers all across the United States. Taking off from a carrier 750 miles from the Japanese coast, Doolittle led a group of 16 B-25 bombers over the Japanese mainland to Tokyo. Because the planes had to sacrifice bombs for fuel, they caused little damage to the capital, but the propaganda and morale value was

SPECIFICATIONS

B-25
Wingspan: 67 ft 7 in (20.59 m)
Length: 51 ft (15.55 m)
Height: 16 ft 4 in (4.98 m)
Wing area: 610 sq ft (56.67 m)
Empty weight: 19,530 lbs (8,858 kg)
Maximum speed: 285 mph (458 km/h) at 15,000 ft (4,572 m)
Service ceiling: 24,200 ft (7,376 m)
Normal range: 1,350 miles (2,172 km) with 3,000 lbs (1,360 kg) of bombs
Maximum range: 2,200 miles (3,540 km) with ferry tanks
Engine: Two R-2600-29 Wright "Cyclone" 14-cyl, air-cooled radial engines developing 1,700 hp each for takeoff
Armament: Eighteen .50-cal (12.7 mm) M-2 Colt-Browning machine guns. Up to 3,200 lbs (1,451 kg) of bombs

incalculable. Since there was no way to land such a large aircraft on a carrier, the planes had to be ditched in China, with most of their crews eventually making their way home safely. Unfortunately, those who did not make it back were captured by the Soviets and Japanese, with the latter executing some of the prisoners.

The Memphis Belle

The B-17 Flying Fortress was known to its crew as the *Memphis Belle*,

SPECIFICATIONS

B-17 FLYING FORTRESS

Wingspan: 103 ft 9 in (31.6 m)
Length: 74 ft 4 in (22.6 m)
Height: 19 ft 1 in (5.8 m)
Wing area: 1,420 sq ft (132 sq m)
Empty weight: 36,135 lbs (16,391 kg)
Maximum speed: 287 mph (462 km/h) at 25,000 ft
 (7,625 m)
Cruise speed: 182 mph (293 km/h)
Service ceiling: 35,800 ft (10,850 m)
Normal range: 2,000 miles (3,219 km) with
 6,000-lb (2,722 kg) bomb load at 220 mph
 (352 km/h) at 25,000 ft (7,625 m)
Engine: Four 1,200-hp Wright R-1820-97, 9-cyl,
 air-cooled single-row radial engines
Armament: Thirteen .50-cal. machine guns;
 normal bomb load of 6,000 lbs (2,724 kg) with a
 max of 17,600 lbs (7,983 kg)

"I am well convinced that 'Aerial Navigation'
will form a most prominent feature in the progress of civilization."

—Sir George Cayley, 1804

for the pilot's back-home sweetheart, Margaret Polk. The plane and its 10-man crew had already flown more than 20 combat missions when Ira Eaker (1896–1987), commanding general of the U.S. Eighth Air Force, ordered that a full tour of duty would require 30 missions instead of 25. Fortunately, the order was not retroactive. They only had to fly their final missions before they could go home.

Already known as a lucky plane, the *Memphis Belle* needed all the luck it had and more when it flew its next mission, a raid on a German fighter factory in Bremen. Of all the bombers taking part in the raid, 19 never made it back. But the *Memphis Belle* successfully threaded its way through some 200 anti-aircraft guns and 150 fighters to complete its mission and return to its base. Although in the course of its career, the B-17 had sustained severe damage—much of its tail had been shot off twice, its engines had to be replaced nine times, as well as both wings—not one of its crew suffered a serious injury.

Eaker retired the *Memphis Belle* after it completed its 25th mission—the first B-17 to do so—and sent both plane and crew home to a triumphant tour of the United States to help support the war effort. Today, the legendary bomber awaits restoration at the Air Force Museum in Dayton, Ohio.

The Flying Wing

The advantages of the flying wing were apparent to aircraft designers from the turn of the twentieth century. Franco-British designer José Weiss built and flew hundreds of flying wing models and gliders between 1902 and 1907, and the concept was experimented with by such great names as Handley-Page, Junkers, Lippisch, and Burnelli.

American engineer Jack Northrop (1895–1981) had experimented with flying wings as early as 1929, with his first true model, the N-1M, flown in 1941. His experimental MX-324/334 was America's first rocket-propelled interceptor. Northrop eventually built several full-scale flying wing bombers for the U.S. Air Force, including the piston-powered XB-35 and the jet-propelled YB-49, perhaps two of the most elegant military aircraft ever flown. Unfortunately, these aircraft were fraught with mechanical problems and were not perfected in time to help in the U.S. war effort.

Opposite: Captain Vincent B. Evans tells mechanic Shirley Poe about some of the exploits of the famed four-engined bomber *Memphis Belle*, at Boston Airport, June 28, 1943, during the aircraft's morale-boosting tour of the states.

Below: The Northrop N-1M, an early example of a flying wing.

Aircraft of World War II

MITSUBISHI ZERO MODEL 52
Wingspan: 36 ft 1 in (11 m)
Length: 29 ft 11 in (9.12 m)
Height: 11 ft 6 in (3.51 m)
Empty weight: 4,136 lbs (1876 kg)
Engine: 1,130-hp NK1F Sakae 21

P-51 MUSTANG
Wingspan: 37 ft .5 in (11.3 m)
Length: 32 ft 2.5 in (9.82 m)
Height: 13 ft 8 in (4.17 m)
Empty weight: 7,125 lbs (3,232 kg)
Operational: 11,600 lb (5,266 kg)
Maximum speed: 437 mph (703 km/h)
Service ceiling: 41,900 ft (12,780 m)
Range: 1,300 miles (2,092 km)
Engine: Rolls-Royce–Packard-built V-1650-7 Merlin 1,590-hp, 12-cyl vee engine
Armament: Six .50-cal. machine guns, external bomb load of 2,000 lbs (908 kg)
  or drop tanks

GRUMMAN HELLCAT
Wingspan: 42 ft 10 in (13.05 m)
Length: 33 ft 10 in (10.31 m)
Height: 14 ft 5 in (4.39 m)
Wing area: 334 sq ft (31 sq m)
Empty weight: 9,060 lbs (4,110 kg)
Performance: Maximum speed: 380 mph
  (612 km/h) at 23,400 ft (7,132 m)

Maximum range: 1,530 mi (2,462 km)
Engine: Pratt & Whitney R-2800-10W
  "Double Wasp" air-cooled radial 2,000 hp
Armament: Six .50-cal. (12.7 mm) Brown-
  ing M-2 machine guns with 2,400 rounds
  mounted in the wings. Later models could
  substitute two 20-mm guns for the two
  inboard .50-cal.

F4F WILDCAT
Wingspan: 38 ft (11.58 m)
Length: 28 ft 9 in (8.77 m)
Height: 9 ft 2.5 in (4.52 m)
Empty weight: 5,758 lbs (2,612 kg)
Engine: 1,200-hp Pratt & Whitney R-1830-86

DE HAVILLAND MOSQUITO
Wingspan: 54 ft 2 in (16.5 m)
Length: 41 ft 2 in (12.54 m)
Height: 15 ft 3 in (4.64 m)
Empty weight: 15,970 lbs (7,243 kg)
   normal: 20,600 lb (9,344 kg);
   maximum gross: 21,750 lbs (9,865 kg)
Maximum speed: 378 mph (608 km/h) at 13,200
   ft (4,023 m)

Service ceiling: 28,000 ft (8,534 m)
Range: 1,400 miles (2,253 km) with 453 Imp. gal.
Engine: Two Rolls-Royce Merlin 25 12-cyl
   60 vee liquid-cooled engines each providing
   1,620 hp at takeoff and 1,500 hp at
   9,500 ft (2,895 m)
Armament: Four 20-mm British Hispano cannons

B-29
Wingspan: 141 ft 3 in (43.05 m)
Length: 99 ft 0 in (30.17 m)
Height: 29 ft 7 in (9.02 m)
Wing area: 1,736 sq ft (529.13 sq m)
Empty weight: 72,208 lbs (32,752 kg)
Maximum speed: 399 mph (642 km/h) at 30,000 ft (9,144 m)
Normal range: 4,200 miles (6,759 km)
Engine: Four Wright Aeronautical R-3350-57 twin row radial 2,200 hp
Armament: Eight or twelve .50-cal. machine guns. One 20-mm cannon
Maximum bomb load: 20,000 lbs (9,071 kg)

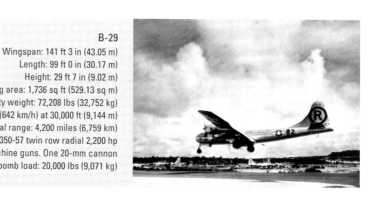

The Korean War

Lt. Col. Bruce Hinton, a pilot with the Fifth Air Force's 4th Fighter-Interceptor Group, smiles broadly as he indicates that he has damaged two enemy MiG jet fighters on his most recent mission on May 1, 1951. In the encounter, four enemy fighters were heavily damaged, with one listed as probably destroyed, with no damage or losses to the American F-86 Sabre jets.

The Korean War saw the first widespread use of jets in combat. The American F-80 saw some service, but no match for the Russian MiG, it was quickly retired and replaced by the North American Aviation F-86 Sabre. The MiG (named for its design team, Mikoyan and Gurevich), first developed in 1947, was at least the equal of the F-86 in most respects, but it suffered from a lack of pilots with experience flying jets. Most of the Sabres were flown by pilots who had experience in the finer points of jet combat. One of the lessons learned from the war was that the new era of jet fighters required not only new flying techniques but new equipment and technology as well.

The Supersonic Fighters

Speed became the prime requisite of military jet aircraft following the Korean War, as well as the development of new technologies that greatly improved the fighting capability of jet aircraft. The introduction of missiles such as the AIM-9 Sidewinder and AIM-7 Sparrow moved combat beyond visual range, requiring the use of radar to acquire targets. Engineers experimented with a variety of innovations, such as the swept wing, delta wing, variable-geometry wings, and area-ruled fuselages. With the aid of swept wings, the new jet fighters were the first production aircraft to break the sound barrier.

The primary applications of the first supersonic military jets were as fighter-bombers (such as the F-105 and the Sukhoi Su-7), and the interceptor (the British Electric Lightning and the F-104 Starfighter). The interceptor developed from the realization that guided missiles would eventually replace guns altogether. As a result, interceptors were equipped with radar

SPECIFICATIONS

F-86 SABRE
Crew: One
Length: 37 ft 6 in (11.4 m)
Wingspan: 37 ft 1 in (11.2 m)
Height: 14 ft 8 in (4.46 m)
Maximum speed: 685 mph (1,102 km/h)
Armament: Six .50-caliber machine guns and
    eight 5-in (12.7-cm) rockets, or 2,000 lbs
    of bombs (907 kg)

> "To become an ace a fighter must have extraordinary eyesight, strength, and agility, a huntsman's eye, coolness in a pinch, calculated recklessness, a full measure of courage—and occasional luck!"
>
> —Gen. Jimmy Doolittle

and designed to carry a large number of missiles, sacrificing agility in favor of speed and rate of climb.

The fastest fighter planes to ever enter service are the Soviet MiG-25, developed during the 1960s, and the related MiG-31 that followed in the 1970s. The goal was to develop a high-speed interceptor capable of speeds near Mach 3 that could defend the Soviet Union from the threat posed by America's XB-70 Valkyrie bomber and A-12 reconnaissance/attack plane.

High speed is no longer considered essential in fighter design. Most high-performance fighters today, like the F-15 Eagle, Eurofighter Typhoon,

Su-30, and F-22 Raptor, have top speeds between Mach 2 and Mach 2.5. This is considered more than adequate for typical combat scenarios, since the emphasis today is on maneuverability rather than speed.

Left: Two important supersonic jet fighters from the late 1950s: The Convair F-106 Delta Dart (top) was a supersonic all-weather interceptor developed from the Convair F-102A Delta Dagger Interceptor. It entered service with the U.S. Air Force in 1959. The Lockheed F-104 Starfighter (bottom) was a high-performance supersonic interceptor aircraft capable of high speed and climb rates. It entered U.S. Air Force service in 1958.

Below: A captured Russian MiG-15 on display at a U.S. Air Force installation in Okinawa, Japan. One of the earliest production jet fighters, it was originally developed in 1948.

The Vietnam War

The F-4 Phantom entered service in 1958, and although retired from U.S. military service in 1996, it still flies in the air forces of eight nations. The Phantom was first used by the U.S. Navy as an interceptor but was also capable of acting as a ground-support bomber for the U.S. Marine Corps.

Like the Korean War, the Vietnam War saw American jet fighters pitted against Russian-built MiGs—in this case the F-4 Phantom II versus the MiG-21. Both jets could achieve speeds up to twice that of sound and were well-matched opponents. The F-4, however, had the edge on versatility and played a wide variety of roles during the conflict. In addition to participating in air-to-air combat, the F-4 carried bombs and provided support for ground troops.

The F-4 Phantom II broke 16 world records during its long career. Among them were a world altitude record of 98,557 feet (30,040 m), set in December 1959 by a prototype XF4H-1, and a world speed record of 1,606 miles per hour (2,585 km/h), set in November 1961.

The B-52

One of the most successful and versatile aircraft ever operated by the U.S. military is the B-52 bomber. Originally designed in the 1950s to deliver nuclear weapons from a high altitude (a role made famous in *Dr. Strangelove*), the eight-engine jet bomber fulfilled a critical function in the Vietnam War, where it was used for low-level saturation bombing using conventional explosives. In a single offensive lasting only one week, B-52s dropped 15,000 tons of bombs on Hanoi and Haiphong. The bomber was still in use, nearly 50 years after its introduction, in the Persian Gulf War, and the U.S. Air Force claims that it will still be in use in 2040!

SPECIFICATIONS

B-52
Crew: Five (pilot, copilot, radar navigator/ bombardier, navigator, and electronic warfare officer)
Length: 159 ft 4 in (48.5 m)
Wingspan: 185 ft 0 in (56.4 m)
Height: 40 ft 8 in (12.4 m)
Empty weight: 185,000 lbs (83,250 kg)
Engines: Eight Pratt & Whitney TF33-P-3/103 turbofans
Maximum speed: 560 knots (650 mph/ 1,000 km/h)
Combat radius: 4,480 miles (3,890 nm, 7,210 km)
Armament: All models up to the H had a pod of four .50-caliber guns that can be loaded with armor-piercing/incendiary ammunition. The H model had one six-barrel 20-mm Vulcan Gatling cannon. Currently the tail guns have been removed on all operating B-52s.
Ordnance: Up to 60,000 lbs (27,000 kg) bombs, missiles, and mines, in various configurations

The Helicopter at War

Although the helicopter had been used extensively throughout the Korean War for rescue operations, reconnaissance, personnel transport, and so on, it was not until Vietnam that the introduction of the gas turbine engine enabled it to become an effective combat aircraft. One of the most familiar images of that war was the formidable Bell AH-1 Cobra gunship. Used throughout the Vietnam War, more than 4,000 were lost. The Cobra was also used extensively in Desert Storm and in the conflict in Iraq.

SPECIFICATIONS

BELL AH-1 COBRA

Crew: Two (pilot, one copilot/gunner)
Length: 44 ft 7 in (13.6 m)
Height: 13 ft 5 in (4.1 m)
Weight: 6,600 lbs (2,993 kg)
Engine: one AVCO Lycoming T53-L-703 turbo-shaft, 1,800 shp (1,300 kW)
Maximum speed: 104 knots (120 mph, 195 km/h)
Armament: Guns: M197 Gatling gun mounted in M97A1 universal turret

Aircraft Carriers

Eugene Ely was the first pilot to take off from a stationary ship in November 1910 when he launched his airplane from the cruiser USS *Birmingham* off the coast of Hampton Roads, Virginia. On January 18, 1911, he reversed this feat to become the first pilot to land on a stationary ship. Ely landed on the aft of the USS *Pennsylvania* using an improvised braking system of sandbags and ropes to stop the plane. His aircraft was then turned around and he was able to take off again.

The first true aircraft carrier was the British HMS *Ark Royal*. A converted merchant ship, it served throughout World War I. The first strike from a carrier against a land target took place on July 19, 1918, when seven Sopwith Camels taking off from the HMS *Furious* attacked a German Zeppelin base. The carrier had no way of recovering the aircraft, however, so the pilots had to either ditch their planes near the ship or land in nearby neutral Denmark.

The Carrier War

Aircraft carriers played a significant role in World War II. In the Atlantic, aircraft taking off from the HMS *Ark Royal* and HMS *Victorious* were responsible for slowing the German battleship *Bismarck* during May 1941. Later in the war, escort carriers proved their worth guarding convoys crossing the Atlantic and Arctic oceans.

The war in the Pacific, however, was truly a carrier war. Japan had entered the war with 10 aircraft carriers, the largest and most modern carrier fleet in the world. The Americans had only six at the beginning of the war, and only three of these were operating in the Pacific. The December 7, 1941, surprise attack on Pearl Harbor was a clear

An Army Air Force B-25B Mitchell bomber preparing to take off from deck of U.S. Navy aircraft carrier *Hornet* during the historic Doolittle raid on Japan in April 1942.

illustration of the power of a large force of modern carriers. In the Battle of the Coral Sea, U.S. and Japanese fleets traded aircraft strikes in the first naval battle in history where neither side's ships were in sight of one another. When four Japanese carriers were sunk by planes from three American carriers at the Battle of Midway, it was considered the turning point of the war in the Pacific. The eclipse of the battleship as the primary component of a fleet was illustrated in 1945 by the sinking of the *Yamato*, the largest battleship ever built, by carrier aircraft.

The Carrier in Korea

In response to the invasion of South Korea, the United Nations command began carrier operations against the North Korean Army on July 3, 1950. This consisted of the U.S. carrier *Valley Forge* and the British *Triumph*. By the end of the conflict, 12 U.S. carriers had served 27 tours in the Sea of Japan. A second carrier unit served as a blockade force in the Yellow Sea. More than 301,000 carrier strikes were flown during the Korean War.

U.S. Carriers in Southeast Asia

Between August 2, 1964, and August 15, 1973, the United States Navy fought "the most protracted, bitter, and costly war" in the history of naval aviation. During that period carrier aircraft supported combat operations in South Vietnam and conducted bombing operations in conjunction with the U.S. Air Force in North Vietnam. Twenty-one aircraft carriers lost 530 aircraft in combat and 329 more in accidents, causing the deaths of 377 naval aviators, with 64 others reported missing and 179 taken as prisoners of war.

View from above the port bow of U.S. Navy aircraft carrier *Saratoga* during flight operations, March 2, 1932. Commissioned in 1927, it was the Navy's third aircraft carrier and saw active service in World War II.

Stealth and Beyond

Familiar aircraft took to the skies during the Gulf War and the war in Iraq, but there were some strange silhouettes, too. More than 100 aircraft took part in the first wave of Desert Storm; huge Airborne Warning and Control System (AWACS) planes scanned the Iraqi borders with their 30-foot rotating radar domes. Airborne tankers began topping off the fighters that had left bases and aircraft carriers far behind. Most of the combat aircraft were conventional enough, and crammed to the gills with the latest, most sophisticated electronic gear: F-14 Tomcats and AH-4 Apache helicopters, for instance. But there was a new type of aircraft the likes of which had not yet been seen.

Weird, angular aircraft that looked more like origami sculptures than like anything that could fly, fifteen F-117A Stealth fighters zipped through enemy radar as undetectably as needles through a window screen. After a century of flying machines that had become ever more sleek and streamlined, the boxy Stealth fighters looked more like something out of Jules Verne than from the last years of the twentieth century. Silent and virtually invisible to radar and other means of detection, stealth aircraft are at the cutting edge of modern military aviation. It is ironic, therefore, that they have their roots in technology that dates from before World War II. The rise in use of effective surface-to-air missiles against aircraft caused

SPECIFICATIONS

F-117
Crew: One
Length: 63 ft 9 in (20.08 m)
Wingspan: 43 ft 4 in (13.20 m)
Height: 12 ft 9.5 in (3.78 m)
Empty weight: 29,500 lbs (13,380 kg)
Engines: Two General Electric F404-F1D2
  turbofans
Maximum speed: 700 mph (1,130 km/h)
Range: Unlimited (with air-to-air refueling)
Armament: Two internal weapons bays. Bombs:
  BLU-109 hardened penetrator, GBU-10
  Paveway II laser-guided, GBU-27 laser-
  guided. Missiles: AGM-65 Maverick
  air-to-surface, AGM-88 HARM air-to-surface

SPECIFICATIONS

B-2
Crew: Two
Length: 69 ft (20.9 m)
Wingspan: 172 ft (52.12 m)
Height: 17 ft (5.1 m)
Empty weight: 158,000 lbs (71,700 kg)
Engines: Four General Electric F118-GE-100
    turbofans
Maximum speed: Classified, high subsonic
Range: 5,600 nm (6,500 mi/12,000 km)
Service ceiling: 50,000 ft (15,000 m)
Armament: 40,000 lbs (18,000 kg) of 500-lb class
    bombs; 27,000 lbs (12,000 kg) of 750-lb CBU-
    class bombs, 16 Rotary Launcher Assembly
    (RLA) mounted 2,000-lb-class weapons; 16
    RLA mounted B61 or B83 nuclear weapons

American military leaders to wonder if perhaps higher and faster was not the best defense. New missiles could catch up with an aircraft no matter how high or how fast it was flying, so perhaps a slower plane would work better if it were simply harder for the missile's radar detectors to "see."

American aviation pioneer Jack Northrop and Reimar Horton in Germany independently had earlier evolved the idea of the flying wing. They were both correct about the aircraft's efficiency. But neither man's design was very stable, and both designs were difficult to fly without resorting to conventional tail surfaces.

Nonetheless, although the goal was to achieve an airplane that possessed far less drag than conventional aircraft, an unexpected side effect of such a configuration was the difficulty of detecting it by radar. With no large fuselage or vertical tail surfaces to reflect a radar beam, flying wings became the ideal choice when the first stealth aircraft were being developed. Special coatings that help to absorb radar energy, along with flat, angled surfaces that reflected radar away from detectors, add to the stealth's invisibility.

The decision to develop the F-117A was made in 1973, with a contract awarded to the Lockheed Advanced

Development Projects "Skunk Works." The first flight was made only 31 months later. The first plane was delivered in 1982, and operational capability was achieved in October 1983, with the last plane of the contract delivered in the summer of 1990.

The B-2 Stealth bomber has seen service in three separate campaigns—in Kosovo in 1999, over Afghanistan in Operation Enduring Freedom, and in Operation Iraqi Freedom.

Look Ma! No Pilot!

Pilotless aircraft go back to World War I. They were for the most part little more than full-size remote-controlled airplanes used as flying bombs and later as drones for target practice. The Kettering Bug of 1918, for instance, was an aerial torpedo, the equivalent of what today would be called a cruise missile, capable of striking ground targets up to 75 miles from launch point. Lately, however, interest has

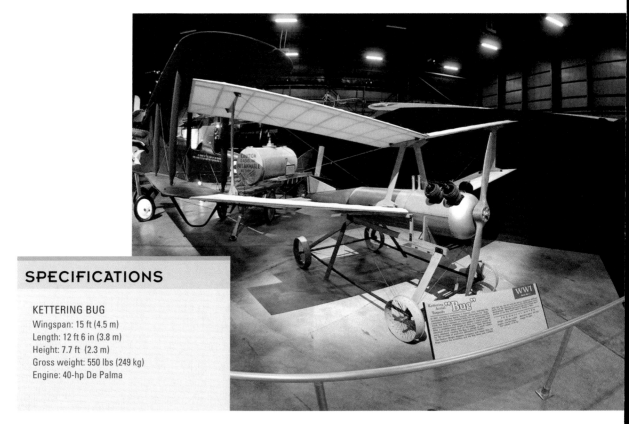

SPECIFICATIONS

KETTERING BUG
Wingspan: 15 ft (4.5 m)
Length: 12 ft 6 in (3.8 m)
Height: 7.7 ft (2.3 m)
Gross weight: 550 lbs (249 kg)
Engine: 40-hp De Palma

SPECIFICATIONS

RQ-2 PIONEER
Length: 14 ft (4.2 m)
Height: 3 ft 4 in (1 m)
Weight: 452 pounds (205 kg)
Wingspan: 16 ft 11 in (5.1 m)
Speed: 110 knots (126 mph / 203 km/h)
Range: Five hours at 100 nautical miles (185 km)
Ceiling: 15,000 feet (4,600 m)
Engine: Sachs 2-stroke, 2-cyl horizontally opposed, simultaneously firing 26-hp

grown within the military since pilotless planes offer the possibility of cheap, efficient combat aircraft that can be used without risk to aircrews. The first Unmanned Air Vehicles (UAV) were used primarily for surveillance. But newer models have been equipped with weapons such as air-to-ground missiles. In the near future, UAVs will undertake bombing and ground attacks. A weapon-carrying UAV is known as an Unmanned Combat Air Vehicle (UCAV).

Israel Aircraft Industries have been pioneering military UAVs since 1982. Their UAV Scout has played an important combat role, though the Israeli Defense Forces uses them mainly for reconnaissance, intelligence gathering, scouting, and communications. The Israeli RQ-2 Pioneer was purchased by the United States and proved its usefulness in the first and second Gulf Wars.

SPECIFICATIONS

GLOBAL HAWK UAV
Crew: None
Wingspan: 116 ft (35 m)
Length: 44 ft (13.4 m)
Height: 15 ft (4.5 m)
Empty weight: 25,600 lbs (11,612 kg) est.
Maximum speed: 454 mph (730 km/h)
Endurance: 42 hours
Engine: Rolls-Royce AE3007 (Allison)

ARMY AIRWAYS EASTERN DIVISION
ARMY AIRWAYS CENTRAL DIVISION
UNITED STATES AIRMAIL
EXTENSIONS TO ARMY AIRWAY

CHAPTER 4

AIRMAIL

WHILE MESSAGE-CARRYING HOMING PIGEONS may have constituted the first airmail, the first human airmail trip occurred in a balloon. On January 7, 1785, Dr. John Jeffries delivered letters to Benjamin Franklin, who was the American minister to France at the time, in a flight from Dover, England, to Calais, France. Balloons also carried messages and mail out of Paris when the city was isolated by Germany during the siege of 1870. The first American airmail was carried by balloonist John Wise, who had taken up ballooning in 1835. The postmaster of Lafayette, Indiana, entrusted Wise with some pamphlets and more than 100 letters, which he was to carry to New York or Philadelphia.

The air was hot and still on the morning of August 17, 1859, and Wise's balloon, the *Jupiter*, failed to reach an altitude where it could find favorable wind. The balloon was blown in the wrong direction, and Wise was eventually forced to land at Crawfordsville, Indiana. From there the mail was transferred to a train. In spite of this inauspicious beginning, the U.S. Postal Service considers this the first official U.S. airmail flight.

Left: Earle Ovington is handed a bag of mail by an unidentified postal worker as New York City postmaster Edward M. Morgan (far left) and U.S. postmaster general Frank Hitchcock look on. Inset: A map showing U.S. airmail routes for the year 1926.

Early Carriers

The invention of the airplane spurred interest in the development of airmail. U.S. postmaster general Frank Hitchcock took an early interest in using aircraft for transporting mail. He authorized local postmasters to send mail by air—as long as it was done at no extra expense to the government. On February 17, 1911, the Petaluma, California, post

Below: The Wiseman-Cooke aircraft, currently on loan to the National Postal Museum in Washington, D.C., is the plane in which Fred Wiseman made his first and only airmail flight.

office had pilot Fred Wiseman (1877–1961) fly 18 miles (29 km) to Santa Rosa. Wiseman had built his own plane based on photos of Curtiss, Wright, and Farman aircraft, as well as from notes he had taken at various air meets. His home-built plane could only fly at 70 mph (113 km/h) and rarely higher than 100 feet (30 m). He carried only three letters, some groceries, and newspapers that he tossed from the air.

Earle Ovington

The first successful mail service with an official U.S. Post Office Department pilot took place at an air show in September of 1911. Earle L. Ovington (1879–1936) had been demonstrating his prowess flying his 70-horsepower Blériot at an air show in New York. On September 23 he was recruited by the U.S. Post Office to carry the first official airmail. Ovington took an oath in Mineola, New York, and became the first official (albeit unpaid) U.S. airmail pilot for the duration of the weeklong show. He delivered 640 letters and 1,280 postcards, including a letter to himself from the United States Post Office designating him as "Official Air Mail Pilot #1." Carrying the bundle of letters between his legs, Ovington flew to an air meet in Mineola, where he dropped the bag of mail from the air. He made similar flights for the remainder of that week.

The Women Airmail Carriers

Although Amelia Earhart is best known for her transatlantic flight and later disappearance over the Pacific, it is not as well known that she often carried mail—albeit unofficially—for the Post Office. These specially canceled letters were intended for stamp

"I happened to be the man on the spot, but any of the rest of the fellows would have done what I did."

—JACK KNIGHT, AFTER THE FIRST NIGHT MAIL FLIGHT

collectors, who avidly sought them, especially first-day deliveries signed by Earhart herself.

Ruth Law bought her first plane in 1912 from Orville Wright himself. An avid aviator, she had tried to persuade President Woodrow Wilson to allow her to fly for the Signal Corps during World War I. After the war, she organized the Ruth Law Flying Circus, in which she demonstrated her skills flying a Curtiss biplane. In 1919 she was awarded a contract to fly official airmail in the Philippines, becoming the first female airmail pilot.

Law once explained the value of women in aviation, saying that women "have qualities which make them good aviators, too. They are courageous, self-possessed, clear-visioned, quick to decide in an emergency, and usually they make wise decisions."

Katherine Stinson (1891–1977) was the fourth American woman to receive a pilot's license. Her brother became a noted aircraft designer, and her sister ran a flight school. During World War I, Katherine flew exhi-

Ben Eielson stands by the cockpit of his De Havilland aircraft in Fairbanks, Alaska.

THE FATHER OF ALASKA'S AIRMAIL SERVICE

Ben Eielson (1897–1929)—who later gained fame for his daring flight over the North Pole—learned to fly during World War I in the U.S. Army Signal Corps' aviation section. He moved to Alaska in 1924, where he worked to establish an airmail route between Fairbanks and McGrath.

bition flights for the Red Cross in her Curtiss JN-4. During this time she also managed to set Canadian distance and endurance records.

In 1918 Stinson began work as an airmail pilot, flying the route between Chicago and New York— several months before the Post Office set up its own official route between the cities. She flew the 783 miles (1,200 km) in just 10 hours, breaking two records in the process.

Katherine Stinson started her career in aviation as an exhibition flier known as the "Flying Schoolgirl."

The First Mail Planes

Above: A combination of out-of-work military pilots and easily available surplus aircraft following World War I created an entirely new breed of pilot: the barnstormer. Traveling from town to town, these daring aviators introduced America to the thrills and wonder of flying.

Right: U.S. mail is loaded into a De Havilland DH-4 from a postal truck.

The first government aircraft to carry mail on a regular basis were six Curtiss JN-4 "Jennies." On May 15, 1918, the Post Office began scheduled airmail service between New York and Washington, D.C., by way of Philadelphia. During the first three months of operation, the army pilots delivered the mail. On August 12, 1918, the Post Office took over the airmail service, replacing the army personnel with civilian pilots and mechanics and the military aircraft with six planes built specifically for carrying mail.

Purpose-built Planes

While the planes they had were successful, the Post Office knew it needed a specially designed aircraft that could better withstand the harsh conditions imposed on mail planes. The result was an adaptation of the De Havilland DH-4

Encouraged by Ovington's success, the Post Office Department authorized 31 experimental flights at fairs, carnivals, and air meets in more than 16 states in 1911 and 1912. These flights convinced the government that airplanes could carry a payload of mail, and in 1916 Congress authorized funding for further experiments. In 1918 Congress appropriated $100,000 to establish experimental airmail routes. Having no planes of its own, the Post Office Department persuaded the Army Signal Corps to provide the planes and pilots needed to start an airmail service.

SPECIFICATIONS

DOUGLAS M-2
Wingspan: 39 ft 8 in (12.09 m)
Length: 28 ft 11 in (8.81 m)
Height: 10 ft 1 in (3.07 m)
Weight: 2,910 lbs (1,320 kg)

SPECIFICATIONS

PITCAIRN PA-5 MAILWING
Wingspan: 33 ft (10 m)
Length: 21 ft 10.5 in (6.4 m)
Height: 9 ft 3.5 in (2.8 m)
Empty weight: 1,612 lbs (731 kg)
Engine: Wright Whirlwind J-5-C, 200 hp

that had seen good service during World War I. The planes, which were designed in Britain and built in the United States, were much more rugged than the JN-4s, but better planes were desperately needed.

In 1926 Western Air Service Inc. began operating the airmail route between Los Angeles and Salt Lake City, via Las Vegas. The approximately 660-mile (1,062-km) route was flown by the Douglas M-2 biplane. Designed specifically for carrying mail, with a 400-horsepower Liberty engine, the M-2 carried mail in a compartment 6 feet (1.8 m) long that was located ahead of the pilot. This compartment was lined with reinforced duralumin,

an aluminum alloy, protected by a firewall, and could hold 1,000 pounds (453 kg) of mail. This compartment could also be adapted to carry one or two passengers.

Texas Air Transport first used the Pitcairn Mailwing in 1927. The Mailwing's reputation was really made on the New York–Atlanta route, a 760-mile (1,223 km) trip that it flew in a third of the time it took by rail. The plane was also popular with some of the fledgling airlines that had cropped up. The company running the New York–Atlanta mail route later picked up the Atlanta–Miami route tool. This was the foundation of what was to eventually become Eastern Airlines.

From Parcels to Passengers

Postmaster general Albert Burleson. In the early days of airmail service, the Post Office maintained its own air force to fly the mail.

The Post Office Department was instrumental in developing technology that eventually made flying safer and more efficient for everyone. Here postmaster general Hubert Work and his assistants are inspecting the innovative radio equipment that was installed in airmail planes in 1922.

The Post Office airmail service laid the groundwork for the fledgling American air transport industry. In 1912 postmaster general Albert Burleson made his first request to Congress for $50,000 for airmail service. But Congress did not release the funds until 1916, authorizing the Post Office to contract with private carriers. The government venture was dead before it began. Private companies thought the money was too little to risk their pilots and airplanes, and none came forward to enter the airmail business. Prospects for a practical airmail system depended on direct, sustained federal action, something that would not become a reality until 1918.

The first two legs of a transcontinental route from New York to Cleveland (with a stop at Bellefonte, Pennsylvania) and from Cleveland to Chicago (with a stop at Bryan, Ohio) began operation in 1919. A third leg opened in 1920 from Chicago to Omaha (via Iowa City), and feeder lines were established from St. Louis and Minneapolis to Chicago. On September 8,

1920, the last transcontinental segment was established from Omaha to San Francisco. Mail was still carried by train at night and flown by day, but even so, the new service improved on cross-country all-rail time by 22 hours.

Ceaseless Labors

In 1921 mail was flown both day and night for the first time over the entire distance from San Francisco to New York. Congress appropriated $1,250,000 to expand airmail service. The Post Office Department installed additional landing fields, towers, beacons, searchlights, and boundary

"I didn't start out to chart the skies; it's just no one had done it before."

—ELREY JEPPESEN

markers across the country. It also equipped the planes with illuminated instruments, navigational lights, and parachute flares. On February 2, 1925, Congress passed a law "to encourage commercial aviation and to authorize the Postmaster General to contract for mail service." The Post Office immediately invited bids, and by the end of 1926, 11 out of 12 contracted airmail routes were operating.

The first commercial airmail flight in the United States occurred on February 15, 1926. As commercial airlines took over, the Post Office transferred its lights, airways, and

FINDING THE WAY WITH ELREY JEPPESEN

Although he never flew for the Post Office Department, Elrey Jeppesen (1907–96) was a pilot who shared their concern for accurate navigation. There was no standardized technique for finding one's way from one city to another, other than following known roads and railways. It was frustrating to follow the winding route of a road when a pilot knew he could cover the distance in a straight line. The best information airmail pilots had to work with was written directions provided by the Post Office. In 1930 Jeppesen began taking careful notes on terrain, landmarks, and landing fields. By 1934, these so-called Airway Manuals had grown into a full-scale book business that eventually became an immensely popular series of charts that are still being updated and sold today.

FEDEX

Federal Express, founded in 1971, officially began operations in 1973 with a fleet of 14 Dassault Falcon 20s that connected 25 U.S. cities. FedEx was the first cargo airline to employ jet aircraft. In 1989 FedEx purchased Flying Tigers, an international cargo carrier founded by veterans of the Flying Tigers who had flown for China in the early years of World War II. As of 2006 the FedEx fleet consisted of 471 planes, including 89 DC-10s and 94 Boeing 727s.

radio service to the Department of Commerce. By September 1, 1927, all airmail was carried under contract. Only seven years after the first official airmail flight, 14 million letters and packages were being delivered each year. Once airmail became accepted, the government transferred airmail service to private companies. Representative Clyde Kelly of Pennsylvania sponsored the Contract Air Mail Act of 1925, which was the first major step toward the creation of a private and profitable American airline industry. Private companies could now bid on the feeder routes that supplemented the transcontinental air lanes. These government contracts helped establish and support the fledgling companies that would eventually become the nation's great airlines.

A Risky Job

Of the first four pilots hired in 1918 by the U.S. Post Office Department, three died in crashes. The occupation continued to be treacherous. In 1919 one pilot died for every 115,325 miles flown, and by 1926 more than 30 pilots had died while transporting the mail.

Aéropostale

Developed in the aftermath of World War I, France's Aéropostale was founded by Pierre-Georges Latécoère (1883–1943) to provide, among other things, airmail services. Latécoère had originally envisioned a company that would provide air routes between France and Senegal by way of Spain and Morocco. He also founded lines between Rio de Janeiro and Recife in Brazil. All of these were ultimately combined into the single company, Compagnie générale aéropostale. In 1933, Aéropostale merged with a number of other aviation companies to create Air France.

Jean Mermoz

Jean Mermoz is regarded as a hero both in Argentina and his native France. A veteran World War I pilot, he joined Latécoère's fledgling company as an airmail pilot. The daring aerobatics of his demonstration flight nearly lost him the job before he had it, but he was soon flying North African routes.

In 1927 Latécoère began building his own line of successful aircraft, many of which he flew himself on Aéropostale routes. His ambition, however, was to establish his own airline connecting France with South America (the existing link between Europe and South America was by ship). To test the feasability of this project, Mermoz was employed to make a hazardous flight over the Andes—at night.

"I learned that danger is relative, and that inexperience
can be a magnifying glass."

—CHARLES A. LINDBERGH

In 1933 Mermoz and Antoine de Saint-Exupéry became central figures in the foundation of Aerolineas Argentinas, both flying extremely dangerous missions for the new company. Mermoz was lost when, concerned that the mail he was carrying would arrive late, he took off in a plane with a bad engine, rather than waiting for repairs. Four hours later, the last radio message was received from Jean Mermoz. Neither he nor his plane was ever found.

Saint-Exupéry

Count Antoine Marie Roger de Saint-Exupéry has become an aviation legend on a par with Charles Lindbergh. Originally a military pilot, Exupéry eventually found himself flying mail for Aéropostale on the Toulouse-Dakar route.

An accomplished writer, Saint-Exupéry published his first book, *The Southern Mail*, in 1928, based on his many harrowing adventures flying in Northern Africa. He was eventually appointed director of Aeroposta Argentina, where he continued writing.

During World War II he flew reconnaissance planes. His plane was shot down, but he managed to escape, ultimately arriving in the United States. On his return to Europe, Saint-Exupéry flew for the Free French and the Allies.

On the night of July 31, 1944, he took off on a mission to observe German troops and was never seen again.

Two of Saint-Exupéry's books have never been out of print and have been read and treasured by readers all over the world. *Wind, Sand and Stars* recounts several of the author's adventures as a pilot. The beloved classic *The Little Prince* is a fantasy that evolved in part from a true story, in which Saint-Exupéry crashed in the depths of the Sahara Desert. He suffered from dehydration and hallucinations for four days before being rescued.

Antoine de Saint-Exupéry examines an aviation chart with his wife, Consuela, who was a noted author in her own right.

CHAPTER 5

THE FIRST ACROSS

LOUIS BLÉRIOT HAD FLOWN across the English Channel in 1909, but there were bigger, more challenging, more dangerous bodies of water to cross. Almost from the beginning, airmen eyed the broad expanses of the Atlantic and, for the supremely ambitious (or supremely foolhardy), the Pacific Ocean.

Aviators were not the only ones to realize that such epochal flights would bring fame, fortune, and immortality to the fliers who first made them. Newspaper owners became quickly aware that sponsoring such flights or offering enormous cash prizes could generate millions of dollars in publicity—even if the flights themselves were unsuccessful.

Alfred Harmsworth, Lord Northcliffe (1865–1922), the fabulously wealthy owner of the *Daily Mail*, offered an award of £10,000 to the first pilot to fly across the Atlantic Ocean. What a challenge that was! The shortest distance possible—between Newfoundland and Ireland—was 1,880 miles (about 3,000 km), which was roughly 10 times the distance any aircraft had flown nonstop up to that point. Northcliffe's rules permitted refueling stops, but they had to be made on water, not land. The only way to avoid refueling was to fly nonstop, but this would require two things—a load of fuel heavier than any plane had ever carried and engines that could run continuously for at least 30 hours, nearly 10 times longer than the average for the time.

Left: Charles A. Lindbergh photographed with his plane, the *Spirit of St. Louis*, after his famous nonstop flight from New York to Paris.

The Great Race

Determined pilots and airplane manufacturers were not daunted by the difficult challenges of completing a nonstop transatlantic flight and continued preparations until World War I began. This turned out to be just as well, since the war accelerated the development of aircraft design, technologies, and engines, so that once the war ended the idea of transatlantic flights was not quite so formidable.

The Atlantic was conquered by air in May of 1919, when an NC-4—one of three U.S. Navy aircraft that attempted the flight—arrived in Lisbon, Portugal. The three planes, an NC-1, NC-3, and the NC-4, left Rockaway, New York, on May 8. The first two were forced down into the Atlantic by severe weather.

The NC-1 crew was rescued by a merchant ship, while the NC-3 was turned into the wind to ride out the storm. Sixty-two hours later, they were able to start their engines and taxi into port at Fayal, in the Azores. The NC-4 made one refueling stop in the Azores and continued to Lisbon. Although the London *Daily Mail* prize was still available, the U.S. Navy had declined to enter the flight in the contest.

Daily Mail Prizewinner

A 26-year-old veteran of the Royal Air Force named John Alcock (1892–1919) had not forgotten the *Daily Mail* prize, which had grown to £13,000 with contributions from other benefactors. Alcock persuaded the aircraft manufacturer Vickers Limited to allow him to test their latest airplane, the Vickers Vimy IV, by making a try for the award. By the time he had recruited Arthur Whitten Brown (1886–1948) as his navigator, there were five other contenders for the prize. One mishap followed another until all of Alcock's competitors were eliminated.

SPECIFICATIONS

CURTISS NC-4
Wingspan: 126 ft (38.4 m)
Length: 68 ft 3 in (20.8 m)
Height: 24 ft 5 in (7.44 m)
Empty weight: 16,000 lbs (7,757 kg)
Engines: Four 400-hp Liberty 12A inline piston

Far right: Capt. John Alcock (left) and Lt. Arthur W. Brown

The great, twin-engine Vickers Vimy lifted from the ground near St. John's, Newfoundland, on the morning of June 14, 1919. Cold was the greatest problem during the long flight. The cockpit was unenclosed and, although it offered partial shelter, was bitterly cold, a particular problem for Brown, whose electric jacket warmer failed.

At one point, while flying blind in clouds, Alcock decreased the Vimy's speed and the plane stalled. He was unable to recover until scarcely 100 feet (30 m) above the freezing North Atlantic waves. Later, heavy rains fell that froze into a coating of ice on the plane, threatening to down the aircraft. As Alcock struggled to gain altitude, Brown stood on his seat, trying to keep his frozen instruments clear by hand.

Finally, they touched down on a bog in Ireland. The plane sank into the soft earth and nosed over. But the pilots were safe and, after a flight of 16 hours and 28 minutes, the Atlantic had been conquered.

SPECIFICATIONS

VICKERS VIMY MARK II
Wingspan: 68 ft 1 in (20.75 m)
Length: 43 ft 6.5 in (13.27 m)
Height: 15 ft 7.5 in (4.76 m)
Empty weight: 7,104 lbs (3222 kg)
Engines: Two 360-hp Rolls-Royce Eagle VIII vee

First Airship across the Atlantic

England claimed the honor of the first airship crossing of the Atlantic when, on July 2, 1919, the R-34, a 634-foot dirigible set off on the 3,130-mile (5,000-km) journey from East Fortune, Scotland, to Roosevelt Field, New York. The world-record airship flight was made in 108 hours and 12 minutes. She made the return trip in just 74 hours 56 minutes, beating her own record.

The British airship R-34 left Britain on July 2, 1919, and arrived in the United States with virtually no fuel left. As the landing party had no experience handling large rigid airships, Maj. E. M. Pritchard jumped by parachute, becoming the first person to reach American soil by air from Europe.

Nonstop: New York to Paris

Raymond Orteig, the wealthy French-born New York hotel magnate who sponsored the prize that inspired Lindbergh's flight.

Right: Charles Lindbergh (left) shakes hands with fellow transatlantic pilot Clarence Chamberlin, with Richard Byrd between them.

Far right: The bare-bones cockpit of the *Spirit of St. Louis.*

As Alcock and Brown were waiting for the right conditions to take off from Newfoundland, an obscure New York hotel owner named Raymond Orteig (1870–1939), who had come to the United States as an immigrant from France at the age of 12, offered $25,000 to the first pilot to fly nonstop from either New York to Paris or Paris to New York.

By 1927, eight years after Orteig had first conceived the idea, there were six major contenders for the prize. They ranged from top French World War I ace René Fonck (1894–1953) and his giant Sikorsky S-35 to barnstormer Clarence Chamberlin (1893–1976) and his single-engine Wright-Bellanca monoplane. Fonck's was not the only famous name in the race. Polar explorer Richard Byrd (1888–1957) and France's air ace Charles Nungesser (1892–1927) had both declared their entries, flying the Paris–New York route. The fifth name in the race was Noel Davis, who was sponsored by the American Legion.

Then, sometime in May, an obscure entrant arrived in a single-engine monoplane—an unknown airmail pilot from the American Midwest named Charles A. Lindbergh.

The Spirit of St. Louis

Starting from San Diego, where his sleek, silver, Ryan-built monoplane had been assembled, Charles Lindbergh (1902–1974) had crossed the continent in two hops. He then took off from New York's Roosevelt Field on the misty morning of May 20, 1927, heading for Paris—3,600 miles (1,609 km) away. Lloyd's of London quoted odds of ten to three that he would never be heard from again.

Lindbergh had left college to pursue a career as a barnstormer. He joined the U.S. Army Air Service as a reserve pilot in 1924 and a year later began flying the mail, shuttling between Chicago and St. Louis. After hearing about the Orteig Prize, Lindbergh persuaded a group of St. Louis businessmen to back

"What kind of man would live where there is no daring?
I don't believe in taking foolish chances, but nothing can
be accomplished without taking any chance at all."

—CHARLES A. LINDBERGH

SPECIFICATIONS

RYAN NYP–*SPIRIT OF ST. LOUIS*
Wingspan: 46 ft (14.02 m)
Length: 27 ft 7 in (8.41 m)
Height: 9 ft 10 in (2.99 m)
Empty weight: 2,150 lbs (975 kg)
Engine: Wright Whirlwind J-5-C, 223 hp

his attempt to conquer the Atlantic. The aircraft he had purpose-built for the flight was adapted from a standard Ryan M-2, an airplane very similar to the Wright-Bellanca Columbia being flown by Chamberlin in his bid for the prize. The wingspan was increased by 10 feet (3 m), and accommodations were made for a larger fuel load, with a fuel tank placed between the engine and the cockpit. This eliminated Lindbergh's forward visibility, but it would provide a buffer between the pilot and the engine in case of a crash. The cockpit was equipped with a periscope so Lindbergh could see ahead for take-off and landing.

At 10:31 on the evening of May 21—33 hours and 31 minutes after leaving New York—Lindbergh landed at Le Bourget field in Paris. He had made the flight alone, carrying only five sandwiches and two canteens of water.

WRONG-WAY CORRIGAN

Barnstorming pilot Douglas Corrigan (1907–95) was denied permission by the Bureau of Air Commerce to make a transatlantic flight in 1935 because his rebuilt Curtiss Robin monoplane was deemed unsafe. Corrigan had been one of the mechanics who worked on the *Spirit of St. Louis* and was anxious to emulate the achievement of his hero, but he remained frustrated for several years, as the federal government still denied him permission.

He was, however, approved for a transcontinental flight from New York to California, and on July 17, 1938, he took off in thick fog and headed . . . east. Corrigan later claimed that he was unaware of his "mistake" until he had been in the air for 26 hours. Corrigan landed in Ireland after a 28-hour, 13-minute flight. All he had carried to eat on his flight were two candy bars, two boxes of fig bars, and a quart of water.

Although angry officials cited him for innumerable infractions, the result of his adventure was only to have his license suspended for two weeks. Corrigan had become a national hero. The people of New York greeted his return to the city with a ticker-tape parade that rivaled Lindbergh's.

First Ladies

Above: Beryl Markham had been a bush pilot before becoming the first person to fly the Atlantic Ocean solo from east to west. She described this and her other flying experiences in her critically acclaimed memoir, *West with the Night*.

Above right: The Fokker F.VII flown across the Atlantic by Wilmer Stultz and Louis "Slim" Gordon.

Below: A Percival Vega Seagull similar to the one flown by Beryl Markham.

Within 18 months of discovering the joy of flying in 1930, Beryl Markham (1902–86)—an Englishwoman who had emigrated to British East Africa—had logged a thousand hours and earned a commercial pilot's license. She went on to earn her living as a bush pilot, carrying passengers, mail, and freight in and out of Nairobi. She had flown from East Africa to London four times when she decided to try a transatlantic flight. Unlike most other pilots, who feared the headwinds they would face, Markham decided to fly the route from east to west. Taking off in her little Percival Vega Gull from Abingdon, England, on September 4, 1936, she flew through the night in darkness and rain. As she approached Nova Scotia, her engine failed, forcing her to glide to a landing in a boggy field. Though the plane nosed over and she was injured, Markham had still managed to cross the Atlantic east to west in 21 hours and 25 minutes, nonstop.

Lady Lindy

One of greatest female aviators made her first flight across the Atlantic as a passenger. Accompanying pilot Wilmer Stultz and his mechanic, Louis Gordon, Amelia Earhart (1897–1937) flew from Newfoundland to Wales on June 17–18, 1928, covering over 2,000 miles in 20 hours

G-AEAD

and 40 minutes. Virtually the entire flight was made in rain, snow, and fog, with Stultz only able to glimpse the sea below during the first hour of the journey.

Although lionized from coast to coast as the hero of the flight and dubbed Lady Lindy, Earhart rankled at what she felt was undeserved credit—although it did give women's aviation a badly needed shot in the arm—and became determined to accomplish things on her own in the future. By 1932 she had determined to fly the Atlantic again—solo.

Taking off from Newfoundland on May 19 in her specially adapted

Earhart preparing to leave on her historic solo flight from Hawaii to California in 1935.

Lockheed Vega, Earhart headed for Paris. It was not long before things started to go wrong. Her altimeter failed, she ran into electrical storms, dealt with icy conditions and, at one point, took a spiraling dive that plunged her to within a few hundred feet of the black ocean. A broken weld in the engine manifold threatened to end the flight in disaster. Earhart knew she would never make Paris and settled for a landing in Ireland. In spite of all her difficulties, her flight took only 15 hours and 18 minutes, setting a speed record for the crossing.

Amelia Earhart with Wilmer Stultz (left) and Louis "Slim" Gordon after their successful transatlantic flight.

From Here to There

The Bellanca WB-2 used by Clarence Chamberlin and Charles Levine for their transatlantic flight.

Clarence Chamberlin, failing in his attempt to win the Orteig Prize, made the first nonstop transatlantic flight from New York to Germany scarcely three weeks after Lindbergh's epochal achievement, covering 3,902 miles (6,280 km) in 42 hours and 45 minutes. Along for the ride was the first transatlantic air passenger, Charles Levine.

San Francisco to Hawaii

American pilots Lester Maitland (1899–1990) and Albert Hegenberger (1895–1983) were the first to make a nonstop flight across the Pacific. Leaving Oakland, California, they covered the 2,400 miles (3,860 km) to Honolulu, Hawaii, in 25 hours and 50 minutes. They flew a Fokker tri-motor airplane dubbed

the *Bird of Paradise*, using the revolutionary new system of following a radio beacon beamed toward Hawaii from a station in Oakland. It was the first time this device had been used for long-distance navigation.

England to Australia

The Australian government had offered a prize of £10,000 for the first flight from England to Australia. It was won in 1919 by brothers Ross (1892–1922) and Keith (1890–1955) Smith. Leaving London on November 12, they arrived in Port Darwin 29 days later after making many stops during the 11,500-mile (18,500-km) journey. The monthlong epic was filled with adventure. Heavy mud made takeoff from Surabaya, Java, impossible, so a makeshift runway was constructed from bamboo mats volunteered by the local people. In addition to the

Major Albert Hegenberger (left) and Lester Maitland with their Fokker C-2 *Bird of Paradise*.

£10,000 prize, both brothers received a knighthood for their achievement.

Coast to Coast

The first nonstop coast-to-coast flight was made May 2–3, 1923, when John MacReady (1887–1979), at left below, and Oakley Kelly (1891–1966), right, of the U.S. Army Air Service flew their Fokker T-2 from New York to San Diego, California. The flight took 26 hours and 50 minutes.

The aircraft had been built by the same Anthony Fokker who was renowned for his fighter planes during World War I. Originally a Fokker F-IV, the T-2 had been converted by the U.S. Army to a transport craft and was later modified for its

SPECIFICATIONS

GOSSAMER ALBATROSS
Wingspan: 97 ft 8 in (29.7 m)
Length: 34 ft (10.4 m)
Height: 16 ft (4.8 m)
Engine: One human being

transcontinental flight. The historic flight laid the groundwork for the transcontinental airlines that were soon to follow.

The *Gossamer Condor*

Designer Paul MacCready's *Gossamer Albatross* became the first plane to meet the standards to win the £50,000 Kremer Prize (established in 1959) for human-powered flight by covering a one-kilometer (0.62 miles), figure-8 course. The *Gossamer Condor* went on to win another Kremer Prize (in 1979), this time for the first crossing of the English Channel by a human-powered aircraft. A later MacCready airplane, the solar-powered *Gossamer Penguin*, made the first sustained solar flight, and in 1981 his *Solar Challenger* crossed the English Channel.

SPECIFICATIONS

FOKKER T-2
Wingspan: 79 ft 7 in (24.26 m)
Length: 49 ft 2 in (15 m)
Height: 12 ft 2 in (3.71 m)
Engine: Liberty V12, 420 hp

CHAPTER 6

THE FIRST AROUND

FLYING FROM CONTINENT TO CONTINENT was one thing, but until 1924 no one had yet succeeded in flying entirely around the world, although aviators from five nations had attempted it. The airplane itself was barely 20 years old, and although it had proved itself to have great military value during World War I, it had yet to show that it had any real future as a means of transportation.

In order to demonstrate the potential of the airplane as a long-distance carrier, the U.S. Army Air Service decided to sponsor a round-the-world flight. The year before, an army plane had made the first coast-to-coast flight in the United States.

Left: Flight commander Maj. Frederick L. Martin, assisted by Sgt. Alva L. Harvey, at Clover Field, Santa Monica, California, about to take off in their Douglas World Cruiser *Seattle* at the start of their round-the-world flight. Inset: Howard Hughes's Lockheed Super Vega *Electra*, the plane in which he made his record, 91-hour flight around the world.

Getting There Faster

When the U.S. government commissioned Douglas Aircraft to create a specially designed airplane for the epic journey, the result was four Douglas World Cruisers. These large, open-cockpit biplanes were each powered by a 12-cylinder, 420-horsepower engine. With a crew of two men each, the planes had no radios and only a

Above left: The Douglas World Cruiser DWC-2 Chicago on exhibit in the National Air and Space Museum in Washington, D.C.

Above right: Wiley Post posing in front of his Lockheed Model 5 Vega *Winnie Mae,* one of the most important individual aircraft in the development of modern aviation science.

compass, altimeter, and turn-and-bank indicator by way of navigational instruments. Named the *Chicago, Seattle, Boston,* and *New Orleans,* the planes departed Seattle, Washington, on April 6, 1924. The expedition was plagued almost immediately by bad weather, and the *Seattle* crashed in Alaska. The remaining three planes

continued, making an eventful—but disaster-free—flight until reaching London. After departing England the *Boston* was forced down into the North Atlantic. The crew was rescued, but the plane was damaged beyond repair.

The round-the-world flight was finally completed when the two remaining planes landed in Seattle on September 28, having flown 27,553 miles (44,340 km)—3,000 miles (4,830 km) more than the circumference of the Earth at the equator—in 175 days.

Around the World by Airship

Like the U.S. Army Air Service, the Zeppelin company wanted to prove the commercial value of long-distance aircraft. In this case, however, the aircraft was the giant, state-of-the-art dirigible *Graf Zeppelin.* On April 7, 1929, the airship departed Lakehurst, New Jersey, for a trip around the world—the first ever attempted by a lighter-than-air craft. It carried 20 passengers—who paid up to $9,000 for a ticket—in luxury that was impossible to match by contemporary airplanes. Its captain, Hugo Eckener, commanded a crew of 40. After crossing the Atlantic, the *Graf Zeppelin* refueled at its base in Friedrichshafen, Germany. From there, it continued on nonstop to Tokyo and on to Los Angeles and, finally, Lakehurst, making the 19,500-mile (31, 300 km) journey in just 21 days and 7 hours, a world record.

Around the World with Wiley Post

On June 31, 1931, pilot Wiley Post (1898–1935) and his navigator, Australian Harold Gatty, took off from New York in an attempt to circle the world in record time. After making only 14 stops, Post returned to New York in just 8 days, 15 hours, and 51 minutes.

Over the following year, he overhauled and improved his plane, including the addition of one of the first gyrocompasses, a compass that does not depend on magnetism, ever installed in an aircraft. On July 15, 1933, Post took off alone from New York. This time, after making only 11 stops, he circled the globe in 7 days, 18 hours, and 49 minutes.

SPECIFICATIONS

GRAF ZEPPELIN
Length: 776 ft (236.6 m)
Diameter: 100 ft (30.48 m)
Volume: 3,708,040 cubic ft (105,000 cubic m)
Payload: 60 metric tons
Engines: Five 550-hp Maybach engines burning Blau gas

SPECIFICATIONS

LOCKHEED 5C VEGA
WINNIE MAE
Wingspan: 41 ft (12.49 m)
Length: 27 ft 6 in (8.38 m)
Height: 8 ft 2 in (2.49 m)
Empty weight: 2,595 lbs (1,177 kg)
Engine: Pratt & Whitney Wasp C, 500 hp

The *Winnie Mae*

Howard Hughes

Millionaire movie producer Howard Hughes (1905–76) transformed circumnavigation of the globe from dangerous experimentation to a level of perfection never before achieved. His methodical attention to detail laid the foundation for the development of long-distance commercial passenger aviation.

Howard Hughes was a leader in the advancement of aeronautical science and engineering.

When he landed his gleaming Lockheed 14 plane in New York on July 14, 1938, he had completed circling the world in just 3 days, 19 hours, and 17 minutes. Hughes made the passage from New York to Paris in half the time Lindbergh had 10 years earlier. The virtually uneventful flight demonstrated the potential of transcontinental air travel, especially since the aircraft he used was originally designed as a 12-passenger commercial carrier.

All the Way Around

The first nonstop flight around the world was made in 1949 by U.S. Air Force captain James Gallagher (d. 1985) and a crew of 14 in a B-50 Superfortress called *Lucky Lady II*. They took off from Carswell Air Force Base in Fort Worth, Texas, on February 26. They were later refueled over the Azores, Saudi Arabia, the Philippines, and Hawaii by KB-29 tanker planes. The round-the-world flight was completed on March 2. *Lucky Lady II* had traveled 23,452 miles (377,000 km) in 94 hours and 1 minute at an average speed of 249 miles per hour (400 km/h).

A Pair of Small Planes

When two Air Force Reserve officers— Maj. Clifford Evans and Maj. George Truman—decided to make the first attempt to circle the world in a small, 104-horsepower private plane, they had understandable difficulty in convincing anyone that it could be done. They chose a Piper Cub, a light plane that was developed after World War II in response to the booming interest in private aviation. It was a small, fabric-covered two-seater, powered by a 90-horsepower Lycoming engine. It became one of the best-selling planes in the private aviation market, and many are still flown by enthusiasts today.

The biggest problem facing Evans and Truman was convincing anyone to back their venture. They eventually persuaded the Piper Aircraft Corporation, Lycoming, and others to provide the necessary equipment and supplies, which included two fully equipped secondhand Piper Cubs. The

SPECIFICATIONS

PIPER L-4 Grasshopper
Wingspan: 10.7 m (35 ft 2.5 in)
Length: 6.83 m (22 ft. 4.5 in)
Height: 1.9 m (6 ft 8 in)
Weight: 309 kg (680 lbs) empty

planes were modified by the addition of instruments and radios, extra fuel tanks, and the replacement of the original wood propellers with metal ones. Once they were ready for flight, the Cubs were dubbed *City of Washington* and *City of Angels*.

Takeoff was from New Jersey on August 9, 1947. After four months and 22,500 miles (36,210 km), the planes returned to New Jersey on December 10. The journey had been made without serious incident and marked the first time a round-the-world flight had been made by planes of under 100 horsepower.

The First Woman Around

Geraldine "Jerrie" Mock (b. 1925) accomplished the feat that had eluded Amelia Earhart when she became the first woman pilot to circle the globe solo. Departing Columbus, Ohio, on March 19, 1964, in her 11-year-old Cessna 180—dubbed the *Spirit of Columbus*—she returned 29 days, 11 hours, and 59 minutes later, after flying 23,103 miles (37,000 km), establishing a speed record for planes under 3,858 pounds (1,750 kg). For her accomplishment, she was awarded the Federal Aviation Administration's Exceptional Service Decoration by President Lyndon Johnson.

The four-seat Cessna 180 was one of the most successful light planes developed after the war. It became especially popular with "bush" pilots, who had to contend with rugged terrain and primitive air strips.

SPECIFICATIONS

CESSNA 180
Wingspan: 36 ft (10.9 m)
Length: 26 ft 2 in (7.9 m)
Height: 7 ft 9 in (2.4 m)
Gross weight: 2,550 lbs (1,157 kg)
Engine: Continental O-470-A

Around the World the Hard Way

Until 1965, round-the-world flights had more or less paralleled the equator. The Rockwell polar flight, however, was the first to circle the globe from north to south, passing over both poles. In addition to conducting valuable scientific research, the expedition set eight world speed records for jet transport.

Pilot Fred Austin learned to fly in 1935 and flew for the military in the Aleutian Islands during World War II. He became an airline pilot for Trans World Airlines (TWA) in 1942, and in 1959 served as chief pilot for the company. When Austin came up with the idea of the circumpolar flight, Flying Tiger Airlines loaned him and his partners a new Boeing 707B. The Rockwell-Standard Corporation financed the flight.

Austin was determined that the flight should make some scientific contribution, and the round-the-world, high-altitude cosmic ray measurements taken during the flight have never been equaled. Among the 26 other experiments were studies of clear-air turbulence and the composition of the atmosphere. The flight was also the first

Jeana Yeager and Dick Rutan. Rutan piloted some of the innovative aircraft designed by his brother, Burt.

to demonstrate Litton Corporation's inertial navigation system. Bernt Balchen was aboard as a guest and was allowed to fly over the South Pole as he had done with Admiral Byrd in 1929.

Around the World Nonstop

The first plane to circle the globe non-stop was conceived in 1981 by engineer Paul MacCready (b. 1925). The resulting *Voyager*—designed and built by legendary aeronautical engineer Burt Rutan—was a unique design intended to maximize lift while minimizing weight and fuel consumption. The gliderlike craft featured a twin-boom tail and small forward wings called canards. The 1,489 gallons (5,600 liters) of fuel required for the flight were contained in virtually every available space: in the fuselage, wings, and twin booms—eventually making up three-quarters of the aircraft's total weight. Dick Rutan (b. 1938) and Jeana Yeager (b. 1952) took off on December 14, 1986. Nine days later, 50,000 people greeted them on their return to Edwards Air Force Base in California after circling the world without stopping or refueling.

The Global Flyer

Steve Fossett (b. 1944) made the first solo nonstop flight around the world between February 28 and March 3,

2005. Taking off from Salinas, Kansas, he returned 67 hours, 1 minute, and 10 seconds later. Flying at an average speed of 342 miles per hour (550 km/h), Fossett also set a world record for speed on a nonstop round-the-world flight.

His plane, designed and constructed by Burt Rutan's company, Scaled Composites, was built of strong, lightweight carbon composites and powered by a single jet engine. Eighty-seven percent of the plane's weight at takeoff was fuel.

On February 11, 2006, Fossett set another record when he left Kennedy Space Center in Florida, flew around the world nonstop, and then continued on across the Atlantic for a second time to finally land in England. He had flown 25,766 miles (41,500 km) in 76 hours and 43 minutes.

Fossett had earlier set two records for transcontinental flights. Flying from San Diego, California, to Charleston, South Carolina, in his *Citation X* in 2 hours, 56 minutes, and 20 seconds, he set a speed record for nonsupersonic jets. Returning to San Diego, he flew the same course as copilot for Joe Ritchie, this time in a Piaggio Avanti turboprop. Their transcontinental time of 3 hours, 51 minutes, and 52 seconds set a record for turboprop planes. It was a busy day for Steve Fossett.

The *Breitling Orbiter 3*

Bertrand Piccard (b. 1958) and Brian Jones (b. 1947) took off from Switzerland in their balloon, the *Breitling Orbiter 3*, on March 1, 1999. Nineteen days, 21 hours, and 55 minutes later they landed in the Egyptian desert after flying 28,431 miles (466,000 km) —the first non-stop round-the-world flight in a balloon.

The balloon had evolved from two previous attempts, allowing Piccard and Jones to refine both their equipment and their techniques. One of the latter was to fly at altitudes of up to 36,000 feet (roughly 11,000 m), where steady jet-stream winds carried them at speeds of up to 105 miles per hour (167 km/h).

SPECIFICATIONS

BREITLING ORBITER 3
The gondola is made of Kevlar and carbon composites. The interior is pressurized, with enough air held in reserve to repressurize the cabin four times. Power was supplied via solar panels that kept onboard batteries charged. The cabin was heated by propane burners.
Length: 17 ft 10 in (5.2 m)
Height: 10 ft 3 in (3.1 m)
Empty weight: 4,400 lbs (1,996 kg)
Fuel: Propane

SPECIFICATIONS

VOYAGER
Length: 29 ft 2 in (8.90 m)
Wingspan: 110 ft 8 in (33.76 m)
Height: 10 ft 3 in (3.13 m)
Empty: 2,250 lbs (1,020 kg)
Engines: One Teledyne Continental 0-240 piston engine (forward) and one Teledyne Continental IOL-200 piston engine (aft)

PART 2

BEYOND THE GOLDEN AGE

CHAPTER 7

AIRLINES

IT WAS ALL WELL AND GOOD to fly for fun and adventure, but eventually pilots and airplane manufacturers began to wonder how to go about making a profit in aviation. Was there any money to be made in hauling freight and passengers? Could aircraft compete in speed, safety, and reliability with railroads and steamships? After World War I the military cut its orders to a tiny fraction of its wartime numbers. Manufacturers desperately needed to find a new market. At the time, there was virtually not a single aircraft in the world capable of carrying passengers in anything remotely resembling comfort and safety.

The first airplane passenger was mechanic Charles Furnas, whom Wilbur Wright had taken aloft at Kitty Hawk in the spring of 1908. A few months later, in September 1908, Lt. Frank P. Lahm of the U.S. Army flew with Orville Wright while the latter was making demonstration flights for the army. A few days later a witness to this event, Lt. Thomas E. Selfridge, also became one of Wright's first passengers—and also the first air passenger in the world to meet his death when one of the propellers snapped and the plane plunged to the ground.

Left: Constant activity at airports is now commonplace. The airline industry has come a long way since the first scheduled service was instituted in 1914. Inset: An airline routing center in Central Islip, New York, probably around 1960. Pages 126–127: The Space Shuttle *Columbia* lifts off in January 1990. The Shuttle successfully completed its mission 10 days, 21 hours, and 4.5 million miles (7,242,048 km) later.

The Great Airships

The *Graf Zeppelin* flying over the U.S. Capitol.

Some of the most successful of the early passenger aircraft were not airplanes. They were lighter-than-air ships: enormous, torpedo-shaped, gas-filled balloons with a rigid internal structure. The rigid airship was brought to a state of technological perfection by the work of Count Ferdinand von Zeppelin, who obtained his first patent in 1895. Zeppelin's first experimental flights were successful enough for him to establish a company—Luftschiffbau Zeppelin GmbH—that would develop and commercialize the rigid airship.

The great zeppelin airship *Deutschland* became the world's first commercial passenger aircraft when it flew from its base at Friedrichshafen to Dusseldorf on June 22, 1910, a flight of 300 miles (483 km). Count Zeppelin himself was in command, with 20 passengers aboard on the nine-hour trip. After several more successful flights between different German cities, the airship was wrecked during a storm only five days after its maiden flight.

Over the next four years, Zeppelin's company built and operated a fleet of four new airships, two of them covering 22,655 miles (36,460 km) and carrying 5,577 passengers between them. During World War II, zeppelins were used as the first strategic bombers.

USS *Shenandoah*

The first American-built rigid dirigible was the USS *Shenandoah*, which flew in 1923. The *Shenandoah* was the first airship to be inflated with nonflammable helium, which was still so rare that the *Shenandoah* contained most of the world's reserves. In September 1925 the *Shenandoah* crashed after encountering thunderstorms over Ohio.

The *Graf Zeppelin*

Meanwhile, Germany was building the *Graf Zeppelin*. Initially planned for experimental and demonstration purposes, it also carried passengers and mail. In October 1928 the *Graf* made its first long-range voyage to Lakehurst, New Jersey. The crew was welcomed with a ticker-tape parade in New York and an invitation to the White House.

In August 1929 the *Graf Zeppelin* departed on a complete circumnavigation of the globe. The 21-day, 30,831-mile (49,618-km) voyage began in Lakehurst, New Jersey, at the request of one of the sponsors of the trip—William Randolph Hearst (1863–1951). Away the airship went, across the Atlantic to Friedrichshafen, Germany, where she refueled before traveling to Tokyo. On the next leg of the journey, the first nonstop flight across the Pacific by any aircraft ended in San Francisco before continuing south to Los Angeles. The airship traveled over Chicago and returned to Lakehurst.

SPECIFICATIONS

ZEPPELIN LZ VII *DEUTSCHLAND*
Length: 485 ft 7 in (148 m)
Diameter: 46 ft (14 m)
Volume: 25,243 cubic yards (19,300 cubic meters)
Engines: Three 375-hp engines

It was decided in 1930 to open the first regular transatlantic airship line. Despite the onset of the Depression and growing competition by fixed-wing aircraft, the *Graf Zeppelin* would transport an increasing number of passengers and mail across the ocean every year until 1936, when it was replaced by the *Hindenburg*.

The USS *Shenandoah* moored to the USS *Patoka* in Narragansett Bay, Rhode Island.

The Hindenburg

The LZ-129 *Hindenburg* and her sister ship LZ-130 *Graf Zeppelin II* were the two largest aircraft ever built. The *Hindenburg* was longer than three Boeing 747s placed end to end. It had cabins for 50 passengers (upgraded to 72 in 1937) and a crew of 61. There was even a piano in the lounge and a smoking room that was pressurized to prevent flammable hydrogen gas from leaking in.

During 1936, its first year of commercial operation, the *Hindenburg* flew 191,583 miles (308,323 km), carrying 2,798 passengers and 160 tons (145 metric tons) of freight and mail. In that same year the airship made 17 round trips across the Atlantic Ocean. It also achieved a record in July of that year, crossing the Atlantic twice in 5 days, 19 hours, and 51 minutes.

Below: The *Hindenburg*, just seconds after it burst into flames.

Opposite page, top: The cabin of a Zeppelin airship that carried passengers between German cities. Passenger dirigibles would eventually be able to cross the Atlantic in about a third of the time that it took an ocean liner to travel the same distance.

Opposite page, bottom: A Zeppelin airship entering a hangar at the U.S. Naval Air Station in Lakehurst, New Jersey.

The Fall of the *Hindenburg*

The *Hindenburg* was destroyed by fire while landing at Lakehurst Naval Air Station in New Jersey on May 6, 1937, after making the first transatlantic crossing to the United States in that year. A total of 36 people (about one-third of those on board) perished in the accident. In spite of the impressive safety record of the Zeppelin Company, the Hindenburg accident shattered public faith in airships, marking the end of the giant, passenger carrying rigid airships. The huge aircraft were seeing their last days in any case, whether public opinion had turned against them or not. By the mid-thirties, transport airplanes had evolved to the point where their speed, comfort, reliability, and low cost made them unbeatable competition to the great airships.

In spite of the disasters that befell the *Hindenburg*, *Shenandoah*, and others, zeppelins overall compiled an impressive safety record. The *Graf Zeppelin*, for instance, logged over one million miles (1.6 million km)—including the first circumnavigation of the globe by air—without a single passenger injury. Other commercial airships, such as the British R100 and the American *Los Angeles* also had spotless safety records.

> "There is no noise beyond the distant murmur of the engines and the sigh of the wind . . . the whole atmosphere is one of tranquility and peace."
>
> —*HINDENBURG* ADVERTISING BROCHURE

SPECIFICATIONS

HINDENBURG
Dimensions: 804 by 135 ft (245 by 41.15 m)
Volume: 780,000 cubic ft of hydrogen
  (198,784 cubic m)
Engines: Four 1,200-hp Mercedes-Benz
Top speed: 84 mph (135.18 km/h)
Passengers: 50

The First Airliners

The world's first scheduled airline service began operating on New Year's Day, 1914. It consisted of a Benoist Type XIV seaplane flown by Anthony Jannus (1889–1916) making two round-trip flights a day between St. Petersburg and Tampa, Florida. The 22-mile (35-km) flight took 20 minutes and cost five dollars. (Passengers weighing more than 200 pounds (90 kg) had to pay extra.) Perhaps in large part because it was an open-cockpit plane, the airline was not a success and failed after only a few months.

Tony Jannus (at right) and friends pose with the Benoist flying boat.

Aircraft Transport & Travel, Ltd.

Britain's Aircraft Transport & Travel, Ltd. provided one of Europe's first regularly scheduled passenger services in 1919. Customers were squeezed into the makeshift cabin of a converted De Havilland 4-A bomber, which was soon replaced by the more powerful DH-16 that offered its four passengers the luxuries of windows and heat.

Within a year the company owned several planes and employed 16 pilots. By August 1920 Daimler Airways entered into competition, offering

Royal Dutch Airlines (KLM) fleet

flights from London to Paris in its fleet of new eight-passenger DH-34s, which pampered passengers with a toilet and steward.

KLM

Royal Dutch Airlines (Koninklijke Luchtvaart Maatshappij, or KLM), the world's oldest continuously operating airline, was founded in 1919 and began service in 1920, employing a single DH-19 on its Amsterdam–London route. The early KLM airliners carried no radios, and their instruments were limited to the very basics: airspeed indicator, tachometer, and fuel gauge. The airline soon had a fleet of more sophisticated planes, such as the Fokker F-111, which could carry five passengers in relative comfort. In May 1921 KLM opened the world's first airline ticket office, introduced bus service from cities to its airfields, and built a hotel-restaurant at the airport in Amsterdam.

HP-42 HANNIBAL
Wingspan: 130 ft (39.6 m)
Length: 86 ft 6 in (26.4 m)
Engine: Four Bristol Jupiters, 2,000 hp total

The Great Lawson Airliner

Alfred Lawson (1869–1954) was one of the most colorful characters in the history of airlines. A retired baseball player, Lawson had founded one of the first popular magazines devoted to aviation, called *Fly*, in 1908. He eventually established his own aircraft factory, which operated until 1919.

After closing the Lawson Aircraft Company, its owner was able to find backing for his plans to develop a large transport aircraft. The result, the Lawson C-2 airliner, was completed later that year. The following week it took off from Milwaukee, Wisconsin, for a trip to New York City and Washington, D.C., attracting crowds wherever it landed. Encouraged by this success, Lawson built an even larger plane in 1921.

However, the 34-passenger, three-engine plane crashed in an attempt to take off, and Lawson left the industry again until 1928, when he attempted a 100-passenger design, complete with sleeping berths, which was never completed.

Lawson's C-2 was the only large multiengine plane of its time to have been designed and constructed expressly for the transport of passengers. Powered by two 400-horsepower Liberty engines, it could carry 26 passengers more than 400 miles (644 km). The cost of the huge plane was its downfall: Surplus military bombers were available at half the price and could be converted into passenger-carrying planes.

Alfred Lawson, founder of the Lawson Aircraft Company.

JUNKERS F-13
Wingspan: 58 ft (17.6 m)
Length: 31 ft 5 in (9.6 m)
Engine: Either a BFW 185-hp engine or a
Junkers 195-hp engine

Toward Luxury

One of the most successful airplanes specifically designed for passenger service was the British Airways 10-passenger Handley Page 0/700. This plane was soon replaced by the W/8, a twin-engine biplane capable of carrying a dozen passengers in true luxury, in a cabin decorated with candelabra.

Until 1926 Handley Page airliners dominated the skies. But then Armstrong-Whitworth introduced its great Argosy, a trimotor biplane that carried 20 passengers along with their baggage and a cabin steward. The passengers were treated to individual windows, electric lights, heat, and a toilet. Argosies flew regular air routes all over Europe, India, and Africa until 1933.

The interior of a British Airways Handley Page 0/700, one of the most successful passenger aircraft of its time.

De Havilland responded to the Argosy with the giant HP-42 Hannibal airliner, which made its first flight in November 1930. It was the first four-engine transport ever employed for regular passenger service. Forty passengers were carried in unprecedented safety and comfort. On the two-hour flight between London and Paris, passengers were treated to a four-course meal served in a cabin as big as a railroad car.

The Great Caproni

One of the greatest of the early aircraft designers was Gianni Caproni (1886–1957). Shortly after World War I, he converted one of his four-engine triplane bombers to a passenger transport. Its double-deck cabin—equipped with a bar, lavatory, and seating for 30—was unique for its time.

In the same year Caproni created his Model 60, one of the most extraordinary airliners ever constructed. It was a monster weighing more than 30,000 pounds (13,608 kg), powered by eight 400-horsepower engines, and sporting three sets of triplane wings. Designed to carry 100 passengers, it proved too underpowered for commercial success.

The last of Caproni's giants was the model 90 of 1930. Powered by six 1,000-horsepower engines, the 33,000-

pound (14,968 kg) plane could fly at a respectable 126 miles per hour (203 km/h). Although not a commercial success, it set several world records for carrying capacity.

The Junkers F-13

One of the first German airliners, a converted Zeppelin Staaken bomber,

could carry 25 passengers, but few saw any service. Germany's airline industry hit its stride in 1919, with the introduction of the Junkers F-13, one of the most advanced aircraft of its time. It was the first purpose-built airliner to have an international market—sold and flown all over the world. The use

of internal bracing in the F-13 eliminated the drag produced by struts and guy wires. The plane could cruise at 106 miles per hour (170 km/h) as five passengers rode comfortably inside. A sixth passenger could ride beside the pilot in the open cockpit.

Top: Caproni CA-60 triple triplane flying boat. Built as a cross between a houseboat and a plane, the CA-60 carried 100 passengers.

Bottom: Gianni Caproni, at front with moustache. Caproni's planes were not commercially feasible but set records for carrying capacity.

Flying Ocean Liners

The interior of a Dornier DO-X flying boat.

Seaplanes, or flying boats, had an immediate attraction for intercontinental airline companies. They did not require the construction of special airports and runways, so the planes could land virtually downtown.

Perhaps the most physically impressive flying boat was the massive Dornier DO-X, introduced in 1929. Its wing, at 157 feet (48 m) wide, stretched 28 feet (8.5 m) above the surface of the water. Twelve of the most powerful airplane engines available at the time were carried in six nacelles mounted atop the wing. The DO-X was capable of carrying a hundred passengers. Though gigantic, the great plane still proved underpowered and was therefore a commercial failure.

Greater Heights

The first regularly scheduled airline service between the United States and a foreign nation employed seaplanes. Aeromarine Airways, using converted U.S. Navy HS-2 seaplanes, began flying passengers from Key West to Havana,

Cuba in 1920. The venture proved successful, and Aeromarine was flying passengers from New York within a year. By then it had upgraded its equipment from the HS-2 (which carried only four passengers in open cockpits) to the more advanced F-5-L, which could carry 14 people in an enclosed cabin. Before the company failed in 1923, it had transported 6,000 passengers without mishap.

Igor Sikorsky introduced his S-38 in 1928 as the largest, fastest amphibian in the world. Seating up to 13 passengers in a cabin with leather-covered walls and mahogany trim, the S-38 had a cruising speed of 110 miles per hour (177 km/h).

Consolidated Aircraft introduced the Commodore in 1931. It was the largest commercial seaplane. The 20 passengers sat in any one of three

QUANTAS

The Australian airline QUANTAS (the letters stood for Queensland and Northern Territory Aerial Services) used Imperial Airways Empire S-23 flying boats for its many long-distance overwater routes. Established shortly after the end of World War I, QANTAS began service with a 300-mile route between two Australian towns. Its only plane was a World War I–vintage observation aircraft that could carry a single passenger.

spacious cabins. In 1930 the New York, Rio, and Buenos Aires Air Lines (NYRBA) merged with PAA, which expanded seaplane service to Miami and the West Indies.

Across the Ocean

This success led PAA to encourage Sikorsky to develop an even larger, more advanced amphibian. The S-40 was capable of carrying 40 passengers and a ton of cargo 500 miles (805 km) at a speed of 115 miles per hour (185 km/h).

The Martin Aircraft Company responded to PAA's call for a long-range amphibian with its M-130, which could carry 12 passengers 3,000 miles

SPECIFICATIONS

BOEING 314 CLIPPER
Wingspan: 152 ft (46.3 m)
Length: 106 ft (32.3 m)
Height: 27 ft 7 in (8.4 m)
Cruising speed: 184 mph (296 km/h)
Maximum range: 5,200 miles (8,368 km)
Engines: Four Wright Cyclones, 6,800 hp total

SPECIFICATIONS

SIKORSKY S-40
Wingspan: 114 ft (34.7 m)
Length: 76 ft 8 in (23.4 m)
Height: 23 ft 10 in (7 m)
Empty weight: 21,300 lbs (9,661 kg)
Cruising speed: 117 mph (188 km/h)
Maximum range: 900 miles (1,448 km)
Engines: Two Pratt & Whitney Hornets,
  575 hp each

(4,828 km) nonstop. Although the plane could carry few passengers compared to the S-40, PAA used it in 1936 to introduce its transpacific route. The plane could cover the distance from California to the Philippines in just 60 hours.

The age of the great passenger seaplanes ended with the spectacular Boeing 314 Clipper. The enormous aircraft could carry as many as 74 passengers. It had two decks connected by a spiral staircase. Below were 10 passenger compartments, as well as a dining room and lounge. Even when one of the Clippers was forced down at sea, the strength and seaworthiness of the great airplane kept all its passengers safe until help arrived.

Juan Trippe and Pan Am

Yale graduate Juan Trippe (1899–1981) was fascinated by aviation and began working with New York's Long Island Airways, an air taxi service for wealthy New Yorkers, in 1923. When the venture failed, Trippe and some colleagues pooled their resources and invested in Colonial Air Transport, which traveled the mail route between New York and Boston. Interested in initiating flights to the Caribbean, Trippe created the Aviation Corporation of America. Based in Florida, the company eventually evolved into Pan American Airways. Pan Am's first flight took off on October 28, 1927, traveling from Key West to Havana.

truly global, offering the first passenger flights across the Pacific and, later, across the Atlantic.

During World War II the company flew more than 90 million miles for the government, carrying military personnel and cargo. Since Pan Am's nine Boeing 314 Flying Clippers were the only aircraft that could carry transatlantic payloads, the government commandeered Pan Am's transatlantic aircraft, crews, and operations. Even before the U.S. entered the war, Pan Am had been building and equipping airfields—from the Atlantic coasts of Central and South America to Africa's west coast and Khartoum, which was the jumping-off point to air-supply the war in North Africa. After America's entry into the war, Pan Am Flying Clippers, using these fields, transported military personnel and supplies to the war zone.

Above: Juan Trippe and the globe he often used to check new routes.

Right: A Boeing 707 built for Pan American World Airways.

Global Air Travel

Trippe's genius lay in his ability to pace the growth of his airline with the range of the airliners being made available. His first efforts were island-to-island hops in the Caribbean and into Mexico. As better planes evolved, he extended his routes into Central and South America. When Trippe adopted the Pan Am Clipper flying boats he was able to pioneer routes that were

"If anybody ever flies to the moon, the very next day Trippe will ask the Civil Aeronautics Board to authorize regular service."

—JAMES M. LANDIS

Boeing to produce an airplane much larger than the 707. The result was the Boeing 747. Pan Am was the first customer of the jumbo jet. Trippe's aggressiveness and innovations were instrumental in the development of the airline industry. He had built one of the world's major airlines almost single-handedly, opening routes that made the world accessible to virtually everyone.

New York City's second airport was built on the site of Idlewild golf course. It was dedicated in July 1948 as New York International Airport and then renamed in 1963, after the death of the president, as John F. Kennedy International Airport, or JFK.

Air Travel for All

Trippe was responsible for several innovations in the airline industry, including the tourist-class ticket. By cutting the round-trip fare from New York to London by more than half, traveling by air became more accessible to middle-income families. Trippe quickly recognized the opportunities presented by jet aircraft and ordered several Boeing 707 and McDonnell Douglas DC-8 airplanes.

Pan Am's first jet flight occurred in October 1958 when a Boeing 707 flew from New York's Idlewild Airport (now John F. Kennedy International Airport) to Paris. The new jets allowed Pan Am to introduce lower fares and fly even more passengers. In 1965 Trippe asked

SPECIFICATIONS

SIKORSKY S-38
Wingspan: 71 ft 8 in (21.8 m)
Length: 39 ft 8 in (12.1 m)
Empty weight: 5,800 lbs (2,631 kg)
Cruising speed: 110 mph (177 km/h)
Engines: Two Pratt & Whitney Wasps,
    420 hp each

Postmaster Brown

Postmaster general Walter F. Brown lighting a huge revolving aircraft beacon atop the Wardman Park Hotel in Washington, D.C., as Maj. Clarence M. Young, Gen. William E. Gilmore, USAAC, and Harry Wardman, hotel owner (left to right) look on.

Postmaster general Walter Folger Brown (1869–1961) played an important role in engineering a series of airline mergers from 1929 to 1930; these mergers helped to create a systematic air transportation system that laid the groundwork for the American airline industry.

Before Brown was appointed, private airmail carriers received up to 80 percent of the revenue from airmail postage. In 1926 the Post Office Department changed how the airlines were paid. Instead of being paid for the number of letters carried, they were paid by the pound.

Brown Cleans House

The airlines immediately realized that they could make money by sending thousands of letters stuffed with thick reports, telephone directories, and even spare engine parts to themselves. One airline contractor mailed itself two tons of printed material. The postage cost more than $6,000, but since the airline was paid by the pound, it received $25,000 from the Post Office. Although government-sponsored airmail service had cost American taxpayers $12 million by the time the transport of airmail had been transferred entirely to private carriers in 1927, the expenditure helped establish a basic air transportation system in the United States.

When Brown took office there were 44 small airline companies. Most of them depended on government airmail contracts for their survival. He believed they were unwilling to invest in new equipment and were operating obsolete aircraft, with the result that they were flying with questionable safety margins. Brown's solution was to eliminate competitive bidding for airmail

"A commercial aircraft is a vehicle capable of supporting itself aerodynamically and economically at the same time."

—WILLIAM B. STOUT, DESIGNER OF THE FORD TRI-MOTOR

contracts and award airmail contracts only to large and sufficiently financed companies. In 1930 the U.S. Congress gave Brown virtually absolute power over the airlines.

Making Airlines Profitable

Under Brown's new authority, the Post Office paid the airlines for space on their aircraft rather than actual mail carried. This encouraged the airlines to purchase larger aircraft. Once mail had been loaded, any excess space could be given over to passengers, allowing the airline to make even more money by selling tickets.

Brown believed in large airlines and strongly supported the idea that just one company should be in control of transcontinental airmail routes. Brown had the authority to force smaller companies to merge or die. The ones that survived eventually evolved into many of the airlines we know today, such as United Air Lines and American Airlines.

Brown was not out to establish monopolies. His goal was to create a system of three self-supporting transcontinental airline systems. When Brown entered office, the cost per mile of airmail was $1.10. When he left in 1934, it was 54 cents. In the late 1920s and early 1930s, all commercial airplanes were called mail planes

regardless of what they were carrying. Today, largely due to Brown's efforts, airliners are considered passenger transports first and foremost.

CHARLES LINDBERGH, AIRMAIL PILOT

When the Post Office Department began transferring airmail duties to private contractors, it proved a boon to up-and-coming aviation companies. This directly affected the livelihoods of individual freelance pilots who could now depend on mail contracts for bread-and-butter income.

Charles Lindbergh received his "Lucky Lindy" nickname not from his famous solo flight across the Atlantic, but from his career as an airmail contract pilot. During that time he had twice survived parachuting from a disabled plane, once when running out of fuel in a blinding snowstorm. Floating slowly to the ground, he was horrified to hear the plane's motor restart and see it heading toward him at full speed. It missed him by just 300 feet.

The Friendly Skies

Although a late starter in developing passenger airlines, by 1929 the United States led the world in the number of passengers and freight being flown annually. On July 4 of that year, Transcontinental Air Transport (TAT) introduced the first coast-to-coast passenger service, although the entire journey was not made by air. In the evenings the passengers transferred to a train for the night leg of the trip. In all it took 48 hours to travel from New York to California.

It was a short-lived novelty, however. That same year TAT introduced the first all-air coast-to-coast service, reducing travel time to just 28 hours. The aircraft they chose for this was the fabulous Ford Tri-Motor.

The Ford Tri-Motor

An airplane introduced by the Ford Motor Company in the late 1920s, the Ford Tri-Motor transformed the American airline industry. The rugged, economical, reliable aircraft quickly became standard equipment on many major U.S. airlines as well as in 95 other nations. The planes could carry 12 passengers and their baggage in relative comfort and safety at speeds up to 122 miles per hour (296 km/h). The Tri-Motor was one of the most successful passenger aircraft ever built, with models still flying well into the 1960s.

The Boeing 247D and the DC-3

The innovative Boeing 247D, introduced in 1934, was quickly superseded by the revolutionary series of Douglas Commercial (DC) aircraft, for which it laid much of the groundwork. Although not widely used by the passenger airlines for which it had been designed,

A Douglas DC-3 takes off from Hullavington airfield in Wiltshire, England.

the 247 and 247D made a mark in transcontinental and interconti-nental air races.

The Boeing engineers were conservative and had failed to take full advantage of the technological and engineering advances that were newly available. So when Donald Douglas (1892–1981) was challenged by William John "Jack" Frye (1904–59) of Trans World Airlines (TWA) to create a better plane than the 247D, his engineers employed the latest innovations in materials, aerodynam-ics, and propulsion technologies. The first result was the twin-engine, 12-pas-senger DC-1 in 1933. TWA placed an order for 20 DC-2s, which had more powerful engines and seated 14 passen-gers. The design impressed a number of American and European airlines, and orders poured in from KLM, Polish Airlines (LOT), Swissair, and Iberia.

SPECIFICATIONS

FORD TRI-MOTOR 5-AT-A
Wingspan: 77 ft 10 in (23.5 m)
Length: 49 ft 10 in (14.9 m)
Height: 13 ft 8 in (4.2 m)
Cruising speed: 110 mph (177 kph)
Engines: Three Pratt & Whitney Wasps, 420 hp each

The Gold Standard in Passenger Transport

A team led by chief engineer Arthur E. Raymond (1899–1999) designed the DC-3 as a result of a phone call from C. R. Smith (1899–1990), the chief executive officer for American Airlines. Smith demanded improvements to the DC-2 design and the new plane used all the latest technology available to Douglas engineers. It was a low-wing, cantilever, all-metal monoplane with trailing edge flaps, single elevator and rudder, and retractable landing gear. It had Jack Northrop's innovative multi-cellular wing structure and two cowled radial engines. Although many vari-ants were eventually built, the original

SPECIFICATIONS

BOEING 247D
Wingspan: 74 ft (22.5 m)
Length: 51 ft 7 in (15.7 m)
Height: 12 ft 1.75 in (3.7 m)
Gross weight: 16,805 lbs (7,623 kg)
Engines: Two Pratt & Whitney Wasp S1H1G, 550 hp each

DC-8 airborne laboratory in flight. NASA's Dryden Flight Research Center employs DC-8s to collect data.

design was so satisfactory that the basic specifications were never changed. The resulting plane made its first flight on December 17, 1935, the 32nd anniversary of the Wright brothers' flight. The ability to travel across America with just one refueling stop, plus the comfort and passenger amenities offered by the DC-3 popularized air travel in the United States.

Early U.S. airlines such as United, American, TWA, and Eastern ordered more than 400 of the new planes, paving the way for the modern American air travel industry. More than five times as many passenger miles were flown in 1941 than in 1935 in the United States,

and much of that can be attributed to the popularity of the DC-3.

After the war, thousands of surplus C-47s, the military version of the DC-3, were converted to civil service and became the standard equipment of almost all the world's airlines, remaining in front-line service for many years. The ready availability of ex-military examples of this cheap, easily maintained aircraft helped to start the worldwide, postwar air transport industry.

Even today, 70 years after the first DC-3, there are still small operators with DC-3s in service. The ability to start and land on grass or dirt runways

"It was the first airplane . . .
that could make money just by hauling passengers."

—C. R. SMITH, PRESIDENT OF AMERICAN AIRLINES, REGARDING THE DC-3

also makes it popular in developing countries, where runways are often unpaved. On October 31, 1956, a Navy DC-3, the *Que Sera Sera*, became the first aircraft to land at the South Pole.

ger plane being offered. Its prototype had already set a coast-to-coast speed record, and the production model could carry its 21 passengers at 180 miles per hour (290 km/h). By the end of the decade the DC-3 was the standard for nearly every existing airline, carrying 90 percent of the world's airline traffic. Many DC-3s continued in active service into the twenty-first century.

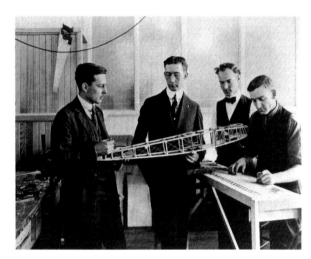

Donald Douglas (left) Glenn Martin (center), and two employees of the Glenn Martin Company check out a model airplane.

Donald Douglas and the DCs

When Donald Douglas introduced his DC-3 in 1935 it immediately changed the course of airline history. Sleek, streamlined, fast, reliable, comfortable, and safe, it was unlike any other passen-

Douglas followed up the successful DC-3 with the 42-passenger DC-4 and the 89-passenger DC-6. The DC-7 Clipper of the late 1950s was the last prop-driven Douglas airliner. It could carry nearly 100 passengers 5,000 miles (roughly 8,000 km), making transatlantic flight a commonplace occurrence. In 1955 Douglas launched its first jet airliner, the DC-8.

SPECIFICATIONS

DOUGLAS DC-3
Length: 64 ft 5 in (19.6 m)
Wingspan: 95 ft (28.9 m)
Height: 16 ft 11 in (4.9 m)
Empty weight: 18,300 lbs (8,300 kg)
Engines: Two Pratt & Whitney 14-cyl Twin Wasps
Cruising speed: 170 mph (274 km/h)
Range: 1,025 miles (1,649 km)

Howard Hughes and TWA

Howard Hughes, in front of his Boeing Army Pursuit Plane.

Howard R. Hughes Jr. (1905–76), one of America's most famous billionaires, was also one of the world's most important aviation innovators. He built one of the key aviation manufacturing companies in history and had major involvement with the growth and fortunes of Trans World Airlines.

Transcontinental Air Transport (TAT), formed on May 16, 1928, was one of the first airlines to try to break away from dependence on U.S. airmail contracts and support itself solely by carrying passengers. TAT offered coast-to-coast service for passengers that combined both air and rail travel. TAT passengers traveling cross-country would ride Pullman railway sleepers at night and fly in Ford Tri-Motors during the day. It worked, but TAT was losing money.

Postmaster general Walter Brown believed that two airlines should not operate over the same route, especially if both were receiving government mail payments. So he suggested that TAT combine its services with Western Air Express, and the two companies merged in 1930, forming Transcontinental and Western Air, Inc. (TWA). The new airline received its first mail contract immediately and began flying coast to coast.

The President and the Stockholder

Jack Frye, a former Hollywood stunt flier and TWA's first director of operations, was instrumental in determining the specifications of the Douglas DC-1 and DC-2 aircraft. In 1934 Frye became president of TWA and made it his job to ensure that TWA was at the forefront of modern technological advances. In 1938, for example, he put in an order for the new 33-passenger Boeing 307 Stratoliner, the first commercial plane with a pressurized passenger cabin. He persuaded millionaire aviator and motion picture producer Howard Hughes to finance this purchase. In 1939 Hughes became the principal stockholder of TWA.

SPECIFICATIONS

LOCKHEED CONSTELLATION
Crew: Five
Capacity: 62-109 passengers
Length: 116 ft 2 in (35.4 m)
Wingspan: 126 ft 2 in (38.4 m)
Empty weight: 79,700 lbs (36,151 kg)
Engines: Four Wright R-3350-DA3 turbo-super-charged radial engines
Cruising speed: 354 mph (570 km/h)
Range: 5,400 miles (8,690 km)

After World War II TWA coveted the transcontinental routes over which American Airlines and United Airlines had been battling. TWA was the most aggressive competitor and put the new Lockheed Constellation into service from New York to Los Angeles in 1946. Although United also introduced transcontinental service on the same day with its DC-4 aircraft, TWA came out the winner, since the Constellation was the better airplane. In 1953 TWA was the first airline to inaugurate regularly scheduled nonstop transcontinental service between Los Angeles and New York.

Until 1946 Pan Am had a monopoly on international service. TWA was one of the first airlines permitted to compete with Pan Am, offering flights to Europe and India.

But after the war TWA's fortunes waned. Despite Howard Hughes's introduction of numerous important innovations—in addition to sponsoring the Constellation, he had made TWA the first airline with both transcontinental and transatlantic routes—his increasing eccentricities eventually cost him his hold on the company, as he was forced to surrender control in 1961.

In spite of its difficulties, TWA continued to remain an important, innovative company. In 1961 it

The TWA terminal at JFK airport, designed by architect Eero Saarinen, opened in 1962. In 2004 it became part of a new Jet Blue terminal.

became the first airline to introduce in-flight movies, and in 1969 it overtook Pan American as the world's number-one transatlantic airline. In February 1970, only one month after Pan Am did so, TWA began flying the Boeing 747 jumbo jet on the New York–Los Angeles route.

THE LOCKHEED CONSTELLATION

The Lockheed Constellation was one of the most distinctive airliners flying from the late 1940s until the advent of jet airliners. With its gracefully curved fuselage and triple tail, it was as beautiful as it was successful. As the first widely used airliner to feature a pressurized cabin, it helped inaugurate affordable and comfortable long-distance air travel for the average person.

The Constellation also has the distinction of setting several records. On April 19, 1944, the second production plane, piloted by Howard Hughes and TWA president Jack Frye, flew from Burbank, California, to Washington, D.C., in just 6 hours and 57 minutes. On the return trip, the aircraft stopped at Wright Field to give Orville Wright his last plane flight, more than 40 years after his historic first flight. He wryly pointed out that the wingspan on the Constellation was longer than the distance of that first flight.

On a Jet Plane

Design work on the first jet-propelled commercial airliner began as far back as 1946. The resulting De Havilland DH-106 Comet made its debut flight in 1949 and—after three years of refinement—was introduced to commercial service in 1952 with its maiden passenger flight from London to Johannesburg, South Africa. The Comet was nearly twice as fast as any propeller-driven passenger plane and in its first year carried more than 30,000 passengers. Trouble began to loom for the radical new aircraft when Comets began mysteriously crashing. Most of these incidents, it was discovered, were due to metal fatigue around the windows. When the metal failed, catastrophic decompression resulted. Although later versions of the Comet fixed this and other problems, and Comets flew until 1997, its damaged reputation never fully recovered.

The Concorde's last flight.

Boeing 707 and the DC-8

The British Comet was the first jet airliner, but the Boeing 707 was the first that was commercially successful. Launched in 1954, it immediately became an extremely popular passenger transport, with Boeing selling 1,010 aircraft before the 707 was retired in 1978.

The disasters that followed the premiere of the British Comet caused Douglas Aircraft to take a cautious approach toward developing its own jet airliner. The DC-8 was customized to the specifications of various airlines. The DC-8 ultimately proved to be an extremely successful aircraft but sold only about half as well as the Boeing plane.

The SST

Beginning in the 1950s the airline industry and the U.S. government (as well as interests in other nations) began thinking about the possibility of supersonic transport—or the SST. The increased speed and potential economies seemed to offset concerns about the need for huge amounts of fuel. The main advantage was that since an SST could fly three times as fast as existing transports, one SST would be able to replace three conventional planes, resulting in great savings in manpower and maintenance.

Conceptual work on the SST occurred simultaneously with the development of supersonic fighter planes. By the early 1960s the designs had progressed to the point where the go-ahead for production was given.

Environmental concerns were strong in the 1960s, and various groups raised issues about the SST, such as

SPECIFICATIONS

DE HAVILLAND COMET 4
Length: 111 ft 6 in (34 m)
Wingspan: 114 ft 10 in (34.7 m)
Height: 29 ft 6 in (9 m)
Empty weight: 75,400 lbs (34,201 kg)
Crew: 4
Maximum speed: 500 mph (804 kph)
Range: 3,225 miles (5,190 km)
Engines: Four Rolls-Royce Avon Mk 524 turbojets

SPECIFICATIONS

DOUGLAS DC-8-32
Crew: Three
Capacity: 124–176 passengers
Length: 150 ft 6 in (46 m)
Wingspan: 142 ft 5 in (43.4 m)
Height: 43 ft 4 in (13 m)
Empty weight: 134,000 lbs (60,781 kg)
Engines: Four Pratt & Whitney JT4A-9 turbojet

the effect of sonic booms and how the plane's exhaust might impact the upper atmosphere. Experiments involving the supersonic B-70 gave some weight to these concerns, and Congress eventually dropped funding for the SST program in 1971.

Concorde 001 made its first test flight from Toulouse, France, in March 1969 and went supersonic in October of that year. The aircraft was now ready for service, but the American public was still in the midst of its opposition to supersonic aircraft flying over land. Since the Concorde had been designed with London–New York flights in mind, this was a devastating economic blow. Flights were allowed into Washington, D.C., however, and the service became so popular that it was not long before the Concorde was flying into JFK.

The popularity of the Concorde started another round of SST studies in the United States, but with jumbo

THE "CONCORDSKI"

The Tu-144 was a supersonic transport constructed under management of the Soviet Tupolev design bureau headed by legendary aircraft designer Alexei Tupolev. Western media nicknamed the plane Concordski, since the Tu-144 was outwardly very similar to the Anglo-French SST (in fact, there were many rumors about industrial espionage having played a substantial role in the design of the Tu-144). A prototype first flew in December 1968 near Moscow, two months before the Concorde. The Tu-144 first broke the sound barrier in June 1969, and in July of that same year it became the first commercial transport to exceed Mach 2.

A Boeing 747-400 operated by the Australian airline Qantas comes in for a landing.

Design work on the first jet-propelled commercial airliner began as far back as 1946. The resulting De Havilland DH-106 Comet made its debut flight in 1949 and—after three years of refinement—was introduced to commercial service in 1952 with its maiden passenger flight from London to Johannesburg, South Africa. The Comet was nearly twice as fast as any propeller-driven passenger plane and in its first year carried more than 30,000 passengers. Trouble began to loom for the radical new aircraft when Comets began mysteriously crashing. Most of these incidents, it was discovered, were due to metal fatigue around the windows. When the metal failed, catastrophic decompression resulted. Although later versions of the Comet fixed this and other problems, and Comets flew until 1997, its damaged reputation never fully recovered.

Boeing 707 and the DC-8

The British Comet was the first jet airliner, but the Boeing 707 was the first that was commercially successful. Launched in 1954, it immediately became an extremely popular passenger transport, with Boeing selling 1,010 aircraft before the 707 was retired in 1978.

The disasters that followed the

THE BOEING 747

- The Boeing 747 has more than 6 million parts made in 33 different countries.
- One of the 747's engines produces more thrust than all four engines of the 707 combined.
- When pressurized, the fuselage contains a ton of air.

premiere of the British Comet caused Douglas Aircraft to take a cautious approach toward developing its own jet airliner. The DC-8 was customized to the specifications of various airlines. The DC-8 ultimately proved to be an extremely successful aircraft but sold only about half as well as the Boeing plane.

The SST

Beginning in the 1950s the airline industry and the U.S. government (as well as interests in other nations) began thinking about the possibility of supersonic transport—or the SST. The increased speed and potential econo-

mies seemed to offset concerns about the need for huge amounts of fuel. The main advantage was that since an SST could fly three times as fast as existing transports, one SST would be able to replace three conventional planes, resulting in great savings in manpower and maintenance.

Conceptual work on the SST

SPECIFICATIONS

AIRBUS A380
Capacity: Up to 853 passengers
Length: 239 ft 6 in (74 m)
Wingspan: 261 ft 10 in (79.5 m)
Height: 79 ft 1 in (24.1 m)
Empty weight: 610,200 lbs (276,782 kg)
Cruising speed: 0.89 Mach
Range: 5,600 miles (9,012 km)
Engines: Four Rolls-Royce Trent 900 turbofans or
   Engine Alliance GP7200 turbofans

SPECIFICATIONS

BOEING 747-200B
Wingspan: 195 feet 8 inches (59.64 m)
Length: 231 feet 4 inches (70.51 m)
Height: 63 feet 5 inches (19.33 m)
Empty weight: 377,000 lbs (17,1004 kg)
Maximum speed: 602 mph (969 km/h)
Engines: Four Pratt & Whitney JT9D-7FW
   turbofans

CHAPTER 8

EXPLORING THE WORLD

ONE OF THE EARLIEST applications for aircraft was to observe the Earth from above. Before the invention of the balloon, no one had ever seen the planet from any higher than a mountaintop, treetop, or church tower. The advent of aviation allowed people to trace the courses of rivers, map out cities, and chart enemy fortifications. Balloons and airships, and later, airplanes, were impervious to rushing rivers, impenetrable jungles, trackless deserts, and towering mountains. Jules Verne was one of the first to argue the value of aviation for exploration. In his very first novel, *Five Weeks in a Balloon* (1863), the heroes explored the depths of Africa from the air.

The Earth's atmosphere was one of the first realms to be explored. At the end of the nineteenth century, little was known about conditions higher than the altitude of the mountains climbed thus far. The scientist-balloonists who rose into the unexplored territory miles above the surface of the Earth were as courageous as the first astronauts.

Left: The airship *Norge* used in Roald Amundsen's 1926 polar flight, at Kings Bay, Spitzbergen, Norway. Inset: Cdr. Richard Byrd (left) and his pilot Floyd Bennet prepare to embark on Byrd's first Arctic expedition.

Scientists in the Sky

Illustration showing Henry Coxwell and James Glaisher aloft in a balloon. The pair's first flight took them to a record altitude of 24,000 feet (7,315 m).

Early explorers sometimes met a tragic end. While Henry Coxwell (1819–1900) and James Glaisher (1809–1903) did manage to survive in 1862 when their balloon reached 30,000 feet (9,144 m)—a height above a significant portion of the Earth's atmosphere—Théodore Sivel (1834–75) and Joseph Croce-Spinelli (1845–75) fared less well, dying from oxygen deprivation on their April 1875 ascent in the balloon *Zénith* with balloonist Gaston Tissandier. Arthur Berson (1860–1943) and Reinard Süring of the Prussian Meteorological Institute were the last men of the nineteenth century to dare altitudes over 30,000 feet when in 1901 they ascended to 35,500 feet (10,820 m), a record that stood until 1931.

Capt. Hawthorne C. Gray of the U.S. Army Air Corps set a U.S. altitude record of 29,000 feet (8,839 m) during his very first balloon flight in 1927. Later that year, Capt. Gray reached 42,000 feet (12,801 m) during his third attempt. Sadly, he ran out of oxygen on the descent and was found dead upon recovery of his balloon. This high-altitude flight was the last attempted in

August Piccard and Paul Kipfer leaving the ground in a balloon equipped with a pressurized gondola.

an open basket until 1955, when newly invented high-altitude pressure suits and spacesuits were first tested.

Into the Stratosphere

Swiss balloonist Auguste Piccard (1884–1962) was the driving force behind the numerous stratosphere balloon flights made during the 1930s. His investigations into cosmic rays required that he rise above the atmosphere in order to study the mysterious radiation. To do this he designed a sealed, pressurized gondola. The result was a 300-pound (136 kg) metal sphere 82 inches (208 cm) in diameter. It could keep two people alive for up to 10 hours above 40,000 feet (12,192 m). Piccard and his assistant, Paul Kipfer, reached an altitude of 51,783 feet (15,787 m) on May 27, 1931. A year later, Piccard and physicist Max Cosyns (d. 1998) reached an altitude of 53,152 feet (16,201 m). The Soviet Union, inspired

"As I stand out here in the wonders of the unknown . . . I realize there's a fundamental truth to our nature. Man must explore . . . and this is exploration at its greatest."

—DAVE SCOTT, APOLLO 15

by these achievements, engineered the largest balloon built up to that time. With an envelope that held 860,000 cubic feet of air, it reached a record 60,700 feet (18,501 m) on September 30, 1933. At the Chicago World's Fair (1933–34), the balloon *Century of Progress*, piloted by Jean (1884–1963) and Jeannette (1895–1981) Piccard, Auguste Piccard's twin brother and sister-in-law, reached a record height of 61,000 feet (18,593 m).

Russia's *Osoaviakhim* was a giant stratosphere balloon that flew on January 30, 1934, reaching an altitude of over 72,000 feet (21,945 m). Unfortunately, the gondola became covered with a thick layer of ice during the descent, causing it to tear loose from the balloon. It crashed to Earth, killing its crew. Undaunted by this tragedy, Jeannette Piccard became the first woman to reach the stratosphere later that year in the refurbished *Century of Progress*.

Explorer II reached an altitude of 72,395 feet (22,066 m) on November 10, 1935. Pilots Orvil A. Anderson (d. 1965) and Albert W. Stevens (1886–1949) were able to see the curvature of the Earth. It was a world altitude record that would stand for the next 21 years.

One of the last manned stratosphere balloon flights took place as part of

the U.S. Air Force's Project Stargazer in December 1962. Pilot Joseph Kittinger (b. 1928) and astronomer William C. White rode in a sealed gondola to a height of 82,200 feet (25,055 m), spending more than 18 hours at that height performing observations.

Above: Jeannette Piccard after her flight in the *Century of Progress* in 1934. Piccard was the first licensed female balloon pilot.

Below: Capt. Joseph A. Kittinger (left) and William C. White in the USAF's Project Stargazer balloon.

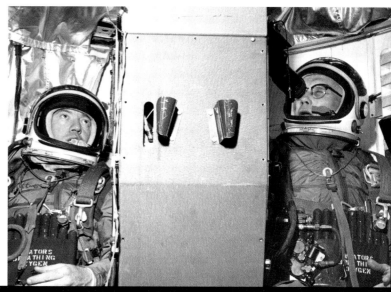

The First to the North Pole

A crew tows a Curtiss Oriole aircraft across a snowy field during Richard Byrd's 1926 expedition.

Col. Umberto Nobile walks inside the dirigible *Norge*, during his historic first flight to the North Pole.

On July 11, 1897, Swedish scientist and engineer Salomon August Andrée (1854–97) set off with two companions for the North Pole in the balloon *Ornen*. Two days later a carrier pigeon arrived with a message saying that all was well. It was the last anyone heard from the balloonists.

The mystery of Andrée's disappearance was not solved until 1930, when the remains of the balloon and its crew were discovered—along with perfectly preserved photographic plates that recorded the expedition's final days. The explorer and his companions had survived the crash of the *Ornen*. They had made camp and were in fact doing very well. What killed them was the abundance of food they had found. The men finally succumbed to poisoning caused by an overdose of the huge amounts of vitamin A in the polar-bear livers they had consumed.

It was not until the year 2000 that a balloonist finally reached the North Pole, when David Hempleman-Adams (b. 1956) achieved this goal in a hot-air balloon.

Byrd's Success

The race to reach the North Pole by air was won by Cdr. Richard Byrd (1888–1957) of the U.S. Navy. Along with pilot Floyd Bennett (1890–1928), he flew over the pole in his specially adapted Fokker trimotor plane on May 9, 1926. Byrd and Bennett covered the 1,600-mile (2,575 km) trip from Spitsbergen, Norway, to the pole in 15 hours and 15 minutes.

Meanwhile, Roald Amundsen (1872–1928), the Norwegian explorer who had been the first to reach the South Pole, made his bid to reach the North Pole by air. Amundsen, though, had decided to go via dirigible. The joint Norwegian-American-Italian venture was led by Amundsen, Lincoln Ellsworth, and Umberto Nobile (1885–1978). The airship *Norge* reached the pole on May 12, just three days after Byrd's epochal flight. Amundsen flew the 2,000 miles (3,219 km) separating Spitsbergen from Point Barrow, Alaska, passed directly over the pole, and continued yet another 700 miles (1,127 km) before bringing the airship to a safe landing in Teller, Alaska. Later, an international dispute erupted between Amundsen and Nobile, who each wanted credit for leading the expedition.

Others Follow

The first transpolar flight by airplane was accomplished by Australians George H. Wilkins (1888–1958), later known as Sir Hubert Wilkins, and Carl Eielson, who had flown the first mail plane in Alaska. They made a 2,200-mile (3,541 km) flight in a Lockheed Vega, from Point Barrow to Spitsbergen, on April 15–21, 1928. One of the goals of the flight was to search for land in the near-polar regions, but they discovered nothing but ice surrounding the pole.

That same year, Umberto Nobile was returning from a successful trip to the pole in the airship *Italia* when the dirigible became covered with ice, broke apart, and was forced down to the icy waters. Badly injured, Nobile survived the crash but was trapped on the ice with nine crewmen. Six others, carried off when the remnant of the dirigible rebounded into the air, were never seen again. The search for the lost aeronauts became an international endeavor, with several nations and more than 20 aircraft participating. Seven survivors were eventually rescued.

RESCUE OF THE ITALIA

When Umberto Nobile's airship *Italia* crashed on its way to the North Pole in May 1928, the international community launched it first polar air-rescue effort. Nobile's colleague and subsequent rival in the *Norge* flight, Roald Amundsen, offered to join in the search, but he lost his life en route to help with the mission. A rescue plane finally reached the stranded crew in June and airlifted Nobile to Ryss Island. When the plane returned for the others it suffered damages, leaving the pilot among those stranded. Eventually the Soviet icebreaker *Krassin* reached the survivors and returned them to safety on July 23, 1928, seven weeks after the crash of the *Italia*.

The flying boat used by Roald Amundsen in his 1925 North Pole expedition returning to Norway.

Antarctic Adventures

Sir Hubert Wilkins and Carl Eielson made the first airplane flight over Antarctica on December 20, 1928. During the 10-hour flight, 100,000 square miles (259,000 sq km) of the frozen continent were observed and mapped.

Richard Byrd, now a rear admiral, made the first airplane flight over the South Pole on November 28–29, 1929. The success of this flight made Byrd the first pilot to fly over *both* poles. With pilot Bernt Balchen and assistants Ashley McKinley and Harold June, Byrd capped an 18-month sojourn at his Antarctic base, dubbed Little America. Arriving at the pole after a flight that had lasted more than 10 hours, the explorers landed and visited a nearby base. They then returned to Little America later that night.

Lincoln Ellsworth and the *Polar Star*

Arctic explorer Lincoln Ellsworth took off from Dundee Island in the Antarctic Weddell Sea on November 23, 1935. He flew in his Northrop Gamma, the *Polar Star*, which he had first taken to Antarctica the year before. After several false starts, Ellsworth and co-pilot Herbert Hollick-Kenyon (1897–1975) finally made their flight across the frozen continent. Flying at

Above: Australian polar explorer, pilot, photographer, and geographer Sir Hubert Wilkins.

Right: A map of the North Pole and the surrounding area indicating the paths of Eielson and Wilkins's two flights over the region. Their unsuccessful March 1927 flight is traced showing the route taken by foot back to Alaska (dotted line).

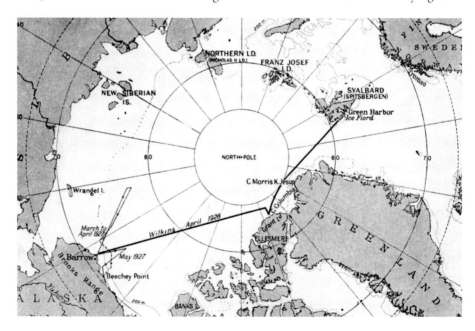

an average altitude of 13,400 feet (4,084 m), they made four landings during the trip. On the final leg of the flight, they ran out of fuel just 25 miles (40 km) shy of their goal: Richard Byrd's Little America, which sat abandoned. It took them six days to walk the remaining distance. They were rescued by a British research ship and later returned to recover the *Polar Star*.

Russians over the Poles

In 1937 Russian aviator Valery Chkalov (1904–38) astonished the United States when he and two crew members flew their single-engine Antonov 25 from Moscow to Vancouver via the North Pole. The 5,500-mile (8,851 km) flight took about 62 hours, setting a new nonstop distance record. The record was broken a month later when another Antonov 25 flew nonstop from Moscow to San Jacinto, California, again via the North Pole. That flight took just over 62 hours.

These flights underscored the growing Russian domination over Arctic flying. By 1948 they had established a scientific base at the pole—the first time any aircraft had actually landed at the North Pole itself. The following year two Russians became the first humans to ever parachute onto the North Pole.

At Home and Abroad

Talbert "Ted" Abrams (1895–1990) was an aviation pioneer and the developer of aerial mapping. As a gunner in World War I, Abrams began taking pictures from airplanes that were used to plan military maneuvers.

After he left the military in 1920, Abrams purchased his first airplane, a Curtiss JN-4 Jenny, and formed his own business, the ABC Airline (Always Be Careful). In 1923 Abrams and his wife, Leota, founded the Abrams Aerial Survey Company. The company created aerial maps of areas on almost every continent, making great strides in map-making and photographic techniques. They photographed the route for US 27 in Michigan, which was the first highway built using aerial photographs.

Abrams formed the Abrams Aircraft Corporation in 1937 to design and build better aircraft for aerial photography. The resulting aircraft, the Model P-1 Explorer, was a low-wing, all-metal monoplane with a pusher engine. Only one of these aircraft was ever built, as attention had turned to World War II. Abrams continued to design other planes, including the Explorer II, but prototypes of these were never built.

Both the Michigan State University planetarium and a mountain in Antarctica were named after Abrams.

Jimmy Angel's Great Discovery

In 1933 bush pilot Jimmy Angel (1899–1956) returned from one of his aerial expeditions into the wilds of Venezuela with the news that he had found the world's highest waterfall. As he passed its brink, he claimed, his altimeter read 6,000 feet (1,929 m). There were few who believed his story.

Returning to what he called the mile-high falls in 1937, Angel managed to land on the summit of the plateau—crashing his Flamingo monoplane in the process. Angel, his wife, and a third party managed to work their way down the mountain on foot, returning to civilization 11 days later. But this time Jimmy Angel had his witnesses.

Below right: Angel Falls in Venezuela, the world's tallest waterfall.

SPECIFICATIONS

ABRAMS P-1 EXPLORER
Crew: Two
Length: 26 ft 6 in (8 m)
Wingspan: 36 ft 8 in (11 m)
Height: 6 ft 4 in (1.9 m)
Empty weight: 2,100 lbs (953 kg)
Engine: Wright R-975 E-1,365 hp
Maximum speed: 225 mph (362 km/h)
Range: 1,400 miles (2,253 km)

Charles Lindbergh inspects damage to the left wing of the *Tingmissartoq*.

What is now known as Angel Falls is indeed the world's highest, with an uninterrupted drop of 3,212 feet (979 m) at an elevation of approximately 6,000 feet (1,829 m). Ironically, the local name for the falls is Devil's Mouth.

The *Tingmissartoq*

The *Tingmissartoq* was a Lockheed Sirius owned by Charles and Anne Morrow Lindbergh that took its name from an Eskimo word meaning "one who flies like a big bird." In 1931 the Lindberghs left Maine for a tour around the North Pacific. They flew across Canada to Nome, Alaska, over Siberia and the Kurile Islands to Japan. From Tokyo they continued on to China, where they landed near Nanking at Lotus Lake, completing the first flight from the West to the East via a northern route. Unfortunately, the aircraft was badly damaged at Hankow and had to be shipped back to the United States.

The next great adventure for the Lindberghs in the *Tingmissartoq* was inspired by the desire of Pan American Airlines to scout and develop new routes between Europe, Greenland, and Newfoundland. Packing the plane with as much emergency equipment and fuel as possible (even its floats contained gasoline), the couple followed the Canadian border to Labrador, after which they made the 650-mile (1,046 km) flight to Greenland, where the plane acquired its name. They crisscrossed Greenland and then, via Iceland, continued on to Europe, where they visited most of the major capitals as far as Moscow. From there the Lindberghs followed the west coast of Africa, crossed the Atlantic to South America, and flew down the Amazon before turning north again. They returned to the United States after passing through Barbados and Trinidad, completing a 30,000-mile (48,280 km) flight that visited four continents and 21 countries. The information they brought back proved invaluable in the creation of commercial air routes across the North and South Atlantic.

SPECIFICATIONS

LOCKHEED SIRIUS *TINGMISSARTOQ*
Wingspan: 42 ft 10 in (13.1 m)
Length: 30 ft (9.1 m)
Height: 14 ft 9 in (4.5 m)
Empty weight: 4,589 lbs (2,082 kg)

CHAPTER 9

FLYING FOR THE FUN OF IT

THRILLING RAPT AUDIENCES WITH daring aerial escapades goes back to the very birth of aviation, when the mere sight of a human being rising above the Earth in a balloon was enough to make hearts go all a-flutter. Then, from the moment the airplane was invented, some intrepid souls entertained audiences with death-defying feats. There was more to these events than sheer thrills, however. In addition to their entertainment value, they helped to introduce and popularize aviation, and the competition among racing rivals helped to accelerate the technological development of aviation.

The establishment of much-publicized trophy races also contributed to the rapid development of aviation technology. Pilots and airplane manufacturers quickly realized the impossibility of adapting standard aircraft for races that demanded high speeds, endurance, and maneuverability. This led to aircraft designed and manufactured from scratch for specific events. Much of this new research and development was later applied to the creation of the great fighter aircraft of World War II.

In recent years amateur pilots have gone back to the thrill of simple aircraft. Ballooning, hang gliding, parasailing, and paragliding utilize minimalist design to provide the thrill of flight for anyone wishing to experience it.

Left: Lincoln Beachey, one of the world's most daring stunt pilots, at the controls of a plane in his signature business suit. Inset: Darryl Greenamyer racing his Lancair Legacy.

Aerobatics and Early Races

Stunt pilot Lincoln Beachey challenges a race car.

Perhaps the first superstar of exhibition flying was Lincoln Beachey (1887–1915), who got his start as a balloonist. Beachey joined the Curtiss Flying School's exhibition team in 1910 and perfected his signature stunt, the death dive—climbing vertically until the plane's engine stalled, then diving back toward the ground and pulling up at the last second. Beachey preferred a specially adapted Curtiss D biplane equipped with extra bracing to make the death dive and other stunts possible. Beachey, who made headlines racing against such famous race car drivers as Barney Oldfield (1878–1946), was killed when the wings came off his plane during an exhibition flight in 1915.

Many features of the Curtiss D made it the perfect choice for a stunt pilot. Not only was it highly maneuverable, but it was easily disassembled for shipping. This same quality also made it easy to replace broken parts. The plane

SPECIFICATIONS

CURTISS D
Wingspan: 26 ft 3 in (8 m)
Length: 25 ft 6 in (7.7 m)
Engine: Curtiss 8-cyl, 60/75/80 hp

also employed bamboo for much of its structure, which gave great strength combined with lightness and safety.

The Jenny

The iconographic airplane of the barnstorming era was surely the Curtiss JN-4, affectionately dubbed the Jenny. It was developed during the final years of World War I, where it saw service primarily as a trainer and ambulance plane; virtually all were declared surplus in 1927. Scores were bought by barnstormers, who traveled around the country taking up passengers for a fee, playing at county fairs, and doing movie stunt work. The rugged design of the plane made it ideal for these purposes and also enabled stunt people to

Below left: Three stunt-people in action: "Bugs" McGowan, Mabel Cowan piloting the plane, and Sig Haugdahl at the wheel of the car, c. 1921.

Below right: A Curtiss Jenny, the aircraft of choice for many barnstormers.

"Racing planes didn't necessarily require courage, but it did demand a certain amount of foolhardiness and a total disregard of one's skin."

—MARY HAIZLIP, PIONEER AIR RACER

leave the passenger cockpit and literally walk all over the plane.

The Schneider Cup

In 1912 wealthy French aviation buff Jacques Schneider created a trophy to be awarded to the winner of an annual seaplane race. In 1925 the race was won by the remarkable Curtiss R3C-2, with an average speed of 232.57 miles per hour (395 km/h) flown by none other than James "Jimmy" Doolittle, who would go on to lead the U.S. air attack on Tokyo in 1942.

Schneider had stipulated that any nation that won the trophy three consecutive times would become the permanent champion. England won that honor in the legendary Schneider Cup race of 1931 with the Supermarine S.6B, an aluminum-clad low-wing monoplane equipped with a powerful 2,350-horsepower Rolls-Royce engine. The race consisted of seven laps around a 31.07-mile (50 km) course. On the first two laps pilot J. H. Boothman averaged 342.9

ORMER LOCKLEAR: WING WALKER

Many people credit Ormer Locklear (1891–1920) with being the first wing walker. Locklear joined the U.S. Army Air Service in 1917 and started climbing out onto his plane's wing while in midair to address mechanical problems. His occasional stunts boosted his colleagues' morale, so Locklear was not reprimanded. After leaving the army in 1919, Locklear became a professional barnstormer known as the King of the Wing Walkers. He became an international star who could make as much as $3,000 per day with his act. He was the first to switch from one plane to another in flight and the first to transfer from a speeding car onto a plane via a rope ladder. Ormer Locklear died during a stunt while working as a Hollywood stuntman just 18 months after turning professional.

miles per hour (552 km/h). A second S.6B, flown by George Stainforth (1899–1942), averaged 379.5 miles per hour (611 km/h) breaking all previous seaplane speed records. Later that year, Stainforth became the first man to exceed 400 miles per hour (644 km/h).

SPECIFICATIONS

CURTISS R3C-1/R3C-2
Wingspans: Upper: 22 ft (6.7 m); lower: 20 ft (6 m)
Length: 20 ft (6 m)
Height: 8 ft 1 in (2.5 m)
Empty weight: 2,150 lbs (975 kg)
Engine: Curtiss V-1400, 610 hp/Curtiss V-1400, 665 hp

James Doolittle poses on a floating Curtiss R3C-2 racer, winner of the 1925 Schneider Cup race.

Racing through the Decades

In addition to the Schneider Cup, a number of other trophy races were established in the 1920s and 1930s. Publisher Joseph Pulitzer (1857–1911) established the Pulitzer Trophy Race in 1920. Originally intended for military aircraft, the race eventually evolved into the Cleveland National Air Race, which was open to all comers. Some of the best aviators of the time were attracted to this race, which was in fact not a single race but rather a series of events that ran for ten days. These ran the gamut from parachute jumping to glider demonstrations to airship racing. In 1929 the race became the home for the Women's Air Derby, the predecessor of the Powder Puff Derby. This race featured top women pilots such as Amelia Earhart, Florence "Pancho" Barnes (1901–75), and Bobbi Trout (1906–2001). The Cleveland National Air Race also became the venue for the Thompson Trophy and Bendix Trophy races. The former was one of the earliest national air races to be established. Created in 1929, it consisted of a 10-mile (16 km) course with pylons that were 50 feet tall (15 m) at either end marking the turns. The Thompson races emphasized high-speed flying at very low altitudes.

The Trophies

One of the most distinctive—if not the most famous—of the Thompson racers was the unique Gee-Bee R-1. Granville Brothers Aircraft of Springfield, Massachusetts, created it specifically for this race. Little more than an engine with wings, the plane boasted several features that were advanced for its time, such as variable-pitch propeller and an air-cooled radial engine.

At the Cleveland National Air Races in 1932, Jimmy Doolittle flew the R-1 to set a new speed record for land planes: 296.28 miles per hour (476.83 km/h) on a 1.9-mile (3 km) straightaway, two days before the official start of the event. During the race itself, Doolittle beat the other seven entrants with a speed of 252.59 miles per hour (406.49 km/h), a record for the Thompson race that was not surpassed for four years.

The Bendix Trophy Race, a transcontinental point-to-point race, was created in 1931 by Vincent Bendix (1882–1945), founder of the Bendix

A brochure for the 1932 National Air Races in Cleveland, Ohio.

Corporation. The last Bendix race was flown in 1964. (It was later revived by AlliedSignal, the company that bought the Bendix brand.) Like the Thompson Trophy, the Bendix attracted many of the top pilots in the country, including Jimmy Doolittle and Amelia Earhart, who was the first woman to enter the race. In fact, women pilots made impressive showings in the early races. Earhart took fifth place in 1935, Louise Thaden (1905–79) won in 1936, and Jacqueline Cochran (d. 1980) won in 1938. The legendary Paul Mantz (d. 1965) was the only pilot to ever win the race three consecutive times, in 1946, 1947, and 1948.

Only one aircraft, the Howard DGA-6, was ever designed specifically for the Bendix race. Designed and built by Ben Howard (1904–70) for the 1935 race and dubbed *Mr. Mulligan,* the plane was unlike almost all the other trophy racers of the time, which tended to be small, single-passenger aircraft. Instead, *Mr. Mulligan* was a big four-seat, high-winged monoplane that included a baggage compartment. The plane, however, had been carefully streamlined and featured a powerful Pratt and Whitney engine. It was designed to fly the race nonstop and at a high altitude, neither of which had been attempted before.

In 1935, Howard beat racing ace Roscoe Turner (1895–1970) by 23.5 seconds with an average speed of 238.7 miles per hour (384.1 km/h). Howard then entered the plane in that year's Thompson race, where, even though it was an entirely unsuited design, it took first place with an average speed of 220.19 miles per hour (354.36 km/h).

Meteor and Gulfhawk

Realizing that his Wedell-Williams racer was becoming outclassed, Roscoe Turner had a new plane designed

HOWARD HUGHES, RACING PILOT

In 1935 Howard Hughes decided to design and build his own racing plane. Setting up shop in a hangar, Hughes and his crew took 18 months to complete the single-seat H-1. There had been nothing quite like it before. Innovations included rivets made flush with the metal skin of the fuselage—Hughes wanted to eliminate the drag that even raised rivets would create—a streamlined engine cowling, and retractable landing gear. Even the tail skid was retractable. The plane also had two sets of interchangeable wings: a short-spanned pair for racing and a pair with a longer span for long-distance flying. Hughes immediately set a new speed record for a land plane when he averaged 352.38 miles per hour (567 km/h) on September 13, 1935.

James Doolittle standing in front of a Granville Brothers' Gee-Bee.

SPECIFICATIONS

GEE-BEE R-1
Wingspan: 25 ft (7.6 m)
Length: 17 ft 9 in (5.4 m)
Maximum speed: 300 mph (482.8 km/h)
Empty weight: 1,840 lbs (834.6 kg)
Engine: Pratt & Whitney Wasp Sr., 800 hp

SPECIFICATIONS

WITTMAN *CHIEF OSHKOSH/BUSTER*
Wingspan: 15 ft 1 in (4.6 m)
Length: 17 ft 5 in (5.3 m)
Gross weight: 500 lbs (226.7 kg)
Engine: Continental, 85 hp

and built. The result was the RT-14 Meteor, which went by various names, depending on Turner's sponsor at the time. In its first competition, the 1937 Thompson Race, Turner placed only third, but he returned the following year to win with a speed of 283.416 miles per hour (456 km/h). It was the second time he had won the Thompson. In 1939 he rocketed past all other planes to achieve an unprecedented third win with a speed of 282.5 miles per hour (454.6 km/h).

One of the most outstanding exhibition aircraft of all time was a plane specially designed for the Gulf Oil Company and Alford Williams (1891–1958), who was head of the company's

aviation department. Similar to the U.S. Navy's F3F fighter, the Gulfhawk II was a single-seat biplane that thrilled audiences from 1936 to its retirement in 1948. It was a featured attraction at air shows, state fairs, and races throughout the United States. In 1938 it made a triumphal tour of Europe, where Williams took the opportunity to display his aerobatic prowess.

The Bearcat

Although designed in 1943 as a fighter for the U.S. Navy, the Grumman F8F Bearcat became better-known as an outstanding racing plane in postwar years. It was the second plane flown by the Blue Angels, the Navy's team of crack exhibition fliers. When the Navy decommissioned the plane in 1958, many were purchased by private pilots. One of these, flown by Mira Slovak, took first place in the inaugural Reno Air Races in 1964. Another captured the first World Air Sports Federation (FAI) time-to-climb record when in 1972 pilot Lyle Shelton took his Bearcat from a standing start to 3,000 feet (914 m) in 1 minute, 31.9 seconds. In 1969 a Bearcat flown by Darryl Greenamyer set the world speed record for a propeller-driven aircraft when his speed of 483 miles per hour (777 km/h) beat the 1939 record set by a Messerschmitt Me.209V1.

SPECIFICATIONS

PITTS SPECIAL
Wingspans: Upper: 17 ft 4 in (5.3 m);
   lower: 15 ft 6 in (4.7 m)
Length: 15 ft 6 in (4.7 m)
Empty weight: 640 lbs (290 kg)
Engine: Usually the Lycoming 100–180 hp,
   but any engine between 85–180 hp

Chief Oshkosh

From 1931 until 1954, Steve Wittman's bright-red, homebuilt racer enjoyed one of the most successful careers in the history of air racing. The miniature, single-seat monoplane was only 17 feet 5 inches (5.3 m) long and fully loaded weighed only 500 pounds (226.7 kg). Named *Chief Oshkosh*, it took third place in its first race in 1931 and the next year won the Glenn Curtiss Trophy with a speed of 166.9 miles per hour (268.6 km/h). In 1947, after a particularly bad crash, Wittman totally overhauled the plane and renamed it *Buster*. That year, Bill Brennand (b. 1924)—a pilot whose small size matched that of the miniature racer—flew *Buster* to victory in the Goodyear Trophy Race with a speed of 165.9 miles per hour (266.9 km/h). The doughty little plane was finally retired in 1954 after its last race (in which it took third place).

The Fabulous Pitts Special

In 1943 Curtis Pitts (1916–2005) built a prototype for what would become known as the most successful aerobatic airplane in history. Pitts's original plane made its debut at air shows in 1946. It was later sold to Betty Skelton (b. 1926), who successfully flew it for years at air shows and competitions as its reputation gradually grew.

The Pitts Special was originally hand-built by Pitts, so only a few were produced each year. In the 1960s Pitts started selling inexpensive detailed drawings so that anyone could build their own Pitts Special. The little plane quickly became a regular sight at air meets not only in the United States but in Europe, South Africa, Canada, and even Jamaica. It was flown by the team members of the U.S. Aerobatic Team in countless competitions and was the plane of choice of other international aerobatic teams as well.

SPECIFICATIONS

GRUMMAN F8F BEARCAT
Wingspan: 35 ft 6 in (10.8 m)
Length: 27 ft 6 in (8.3 m)
Height: 11 ft 9 in (3.6 m)
Maximum speed: 434 mph (698 km/h)
Empty weight: 7,070 lbs (3,207 kg)
Engine: Pratt & Whitney R-2800-34W, 2,100 hp

SPECIFICATIONS

GRUMMAN G-22 GULFHAWK II
Wingspans: Upper: 28 ft 7 in (8.7 m);
    lower: 26 ft 1 in (7.9 m)
Length: 23 ft (7 m)
Height: 10 ft (3 m)
Weight: Cross-country: 4,195 lbs (1,902 kg);
    aerobatic: 3,583 lbs (1,625 kg)
Engine: Wright Cyclone R-1820-G1, 1,000 hp

Ballooning, Gliding, and Sailing

The first hundred years in aviation history focused on the development of lighter-than-air flight. Just a few years after the invention of the hot-air balloon in 1783 and, almost immediately, the gas balloon, literally hundreds of balloonists were soaring above the landscapes of Europe and America.

The first successful heavier-than-air flying machines were gliders, which are aircraft with no engines. All the truly significant developments in aviation up to 1903 and the Wright brothers were accomplished with gliders.

Powered aircraft then became the focus of engineers and aviators, but balloon and glider enthusiasts did not fade away entirely. While amateur airplane pilots sought their thrills from stunts and speed, others were following quieter, but no less exhilarating, aeronautic pursuits.

Ballooning

Balloons are relatively cheap and easy to build, so almost anyone could take to the air. In the nineteenth century, while science benefited greatly from the development of the balloon and its ability to explore the upper atmosphere and allow leisurely observations of the Earth, most balloons were flown either for the sheer excitement of it or by daredevils who hoped to make a living thrilling audiences. By the end of the century, however, ballooning had become an expensive, genteel sport indulged in by wealthy dilettantes in much the same spirit they kept race horses, played polo, or raced yachts.

The first international balloon race was held in 1906, sponsored by James Gordon Bennett (1841–1918), the publisher of the *New York Herald*. The race was held in Paris with 16 entrants.

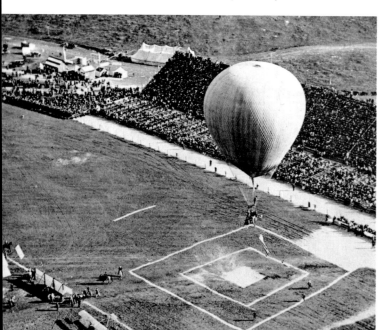

A balloon takes off from the grounds of the 1910 National Balloon Race in Indianapolis, Indiana.

HOW DO BALLOONS GET BACK TO WHERE THEY STARTED?

A group of people on the ground, called a chase crew, follows along in a van or truck. They have radio contact with the pilot, so they can be there to retrieve the balloon when it lands.

> *"Gliders, sailplanes, they are wonderful flying machines. It's the closest you can come to being a bird."*
>
> —Neil Armstrong

The prize, given for the greatest distance flown, went to two U.S. Army officers, Frank P. Lahm (1877–1963) and Henry B. Hersey, who landed in England.

The annual Gordon Bennett Cup races became wildly popular and were held every year until 1939, with a hiatus from 1914 to 1920 because of the war.

The Modern Hot-Air Balloon

Ballooning as a sport was pretty much a dead issue after World War II. Not only did it seem very dull and old-fashioned, but it had also become prohibitively expensive.

In 1955 aeronautical engineer Ed Yost (b. 1919) decided that in order to revive ballooning as a sport it might be necessary to look back to the very beginning of ballooning— to the hot-air balloon. Yost believed that, with the advent of modern materials, engineering, and techniques, the hot-air balloon could be revived.

He concentrated on finding the best materials and shape for the envelope and ways in which the flame could be throttled or even turned on and off at will. He ultimately settled on nylon for the envelope and propane gas for fuel. The gas was cheap, and there were already burners in existence that could be adapted for his use. In 1960 Yost flew the first modern hot-air balloon, and others took notice. The simple elegance of the balloon's design made it the basic pattern for virtually all hot-air balloons to follow.

More than 700 hot-air balloons taking off from the Balloon Fiesta in Albuquerque, New Mexico.

Taras Kiceniuk Jr. takes off in his glider, *Icarus I.*

When Yost and Don Piccard flew a hot-air balloon across the English Channel in 1963, interest in ballooning was revived around the world. In the following year, only six hot-air balloons were registered with the Federal Aviation Administration—by the end of the century, there were more than 10,000.

Ballooning the Atlantic

An early goal of balloonists was to cross the Atlantic. Many attempts were made in the 1800s, and all ended disastrously. When Maxie Anderson (1934–83) and Ben Abruzzo (1930–85) decided to make their own bid for the crossing, they commissioned Ed Yost to construct their balloon, the *Double Eagle I.* Taking off from Massachusetts in 1977, they were forced down into the ocean north of Iceland and nearly froze to death before rescue came.

Joined by a third balloonist, Larry

Newman, Anderson and Abruzzo made another attempt in 1978. The same gondola was used as in the first balloon, but the new envelope could hold 160,000 cubic feet of helium (4,500 cubic meters) compared to the first balloon's capacity of 101,000 cubic feet (2,860 cubic meters). Taking off this time from Maine, the trio landed safely six days later in France. Only four years later, in 1982, the first transatlantic balloon race was held.

Rosie O'Grady

Balloonist Joseph Kittinger (b. 1928) became the first solo pilot to cross the Atlantic in a balloon. On September 14, 1984, Kittinger left Caribou, Maine, in his balloon, the *Rosie O'Grady.* Landing in Italy 83 hours and 40 minutes later, Kittinger had set a new world's long-distance record. Kittinger's balloon, as did most other long-distance balloons, employed the relatively new Rozier technology, which combined the best features of gas and hot-air balloons. Hot air driven into a sealed helium balloon allowed Kittinger to control the altitude of the craft without wasting either gas or ballast.

Gliding

Gliding for the sheer pleasure of it was not really done until after World War I. The Treaty of Versailles had put severe restrictions on the development of powered aircraft in Germany, especially single-seat planes that might potentially evolve into fighters. German aeronautical engineers therefore turned their talents toward perfecting the glider. The first glider meet in Germany was held in 1920, and within 10 years it had become an international event. At the 1936 Olympics in Berlin, gliding was a demonstration sport, and full-sport status was expected for the 1940 games.

A glider flies like any other airplane: Air flowing unevenly over its wings creates lift. But without an engine, how can a glider stay up? What keeps it from slowly spiraling down to the

ACHIEVEMENTS IN GLIDING

The Fédération Aéronautique Internationale (FAI) oversees and authenticates all aviation records. Its Silver Badge is awarded to glider pilots who have reached an altitude of at least 1,000 meters (3,281 ft), remained aloft for at least five hours, and have flown a cross-country distance of at least 50 kilometers (31 miles). There are more than 6,000 glider pilots in the United States alone who hold Silver Badges. Gold and Diamond Badges are awarded for higher and longer flights.

In 1988 Kanellos Kanellopoulos of Greece set the world record for straight distance at 71.5 miles (115.11 km) in his aircraft *Daedalus 88*. This beat the previous record held by Bryan Allen (22.25 miles/35.82 km), which he set with the *Gossamer Albatross* in 1979. Kanellopoulos also set the record for duration in 1988 at 3 hours, 54 minutes, and 59 seconds, again beating the *Albatross*.

ground like a paper airplane? A skillful glider pilot keeps the aircraft aloft for a much longer time than if it were unpiloted. One technique is to look for

Gliders are unpowered aircraft that range in size from paper airplanes to the Space Shuttle, which acts as a glider during reentry and landing.

thermals, great columns of warm air that rise from the ground high into the sky, like inverted, invisible waterfalls of air. By riding these thermals, and sailing from one to another, a glider pilot can stay aloft for hours.

Hang Gliders

Hang gliding harks back to the experiments of Lilienthal and Chanute during the late nineteenth century. Unlike conventional gliders, hang gliders possess no enclosed fuselage. Instead, the pilot of a hang glider is suspended in a harness below the wing. Since it consists of little more than a wing and a pilot, a hang glider is about the simplest flying machine that one could imagine.

Modern hang gliding was strongly influenced by NASA engineer Francis Rogallo (b. 1912). He invented the Flexkite in 1948, a device intended as a landing system for astronauts returning to Earth and seriously considered for use during the Gemini program in the 1960s. After certain improvements, the Flexkite was the standard for wing design for many years.

The first important hang gliders to abandon the Rogallo wing were the *Icarus I* and *Icarus II* biplane designs built in the early 1970s by Taras Kiceniuk Jr. (b. 1955), *Icarus V*, a monoplane version of the previous designs, was the immediate precursor to the modern hang glider. In the late 1990s Exxtacy, the first commercially

A parasail is attached to a long towrope and pulled along by either a jeep or a boat.

successful rigid-wing hang glider, became available.

Paragliding and Parasailing

Paragliding is a recreational and competitive flying sport closely related to hang gliding. A paraglider is a gliding aircraft that is launched by foot. The pilot is suspended below a fabric wing in a harness. The shape of the wing is formed by the pressure of air entering vents in the front. Controls held in the pilot's hands pull down the trailing edge of the wing to control speed and turning.

In parasailing, one or two people are towed behind a vehicle, usually a boat, while attached to a parachute. The parasailer has little or no control over his progress, and therefore parasailers fly strictly for fun and do not compete, as hang gliders and paragliders do.

Land-based parasailing, however, in which a parasail is towed behind a car or snowmobile, has gained traction as a competitive sport in parts of northern Europe.

Hang glider pilots control their aircraft by shifting their body weight.

ODD JOBS

In World War II the U.S. military used enormous gliders to deliver soldiers and equipment to battle zones. The gliders were towed in long trains behind large powered aircraft, and their pilots needed only enough skill to land the disposable aircraft. The entire nose section (including the pilot's compartment) of the Waco CG-4A glider swung upward, allowing access to its cargo compartment. This made it possible to quickly load and unload the glider.

The U-2 spy plane, which became well known in the 1960s after one piloted by Francis Gary Powers (1929–77) was shot down over the former USSR, was essentially a jet-propelled glider. Its gliderlike characteristics enabled it to stay aloft at extremely high altitudes for long periods with a minimal expenditure of energy.

Still in use today, the U-2 provides daily peacetime data and, when requested by agencies such as the Federal Emergency Management Agency, photography supporting disaster relief efforts. Critical intelligence data was provided by U-2s during all phases of Operations Desert Storm and Allied Force.

CHAPTER 10

FLYING HIGHER

HUMAN BEINGS ARE SOMEHOW attracted to high places. They like to climb trees and mountains, or ride Ferris wheels and visit the tops of tall buildings. When the balloon was invented, the thrill came only in part from being able to fly. The rest came from seeing the Earth from a loftier vantage point than anyone had ever experienced.

The early pioneering balloonists ventured into thin air, flying higher and higher, testing the human body's ability to rapidly ascend to high altitudes. Unlike mountaineering, where climbers had plenty of time to acclimatize, the early balloonists experienced sudden changes in altitude. Some of the effects of this were similar to the "bends" deep-sea divers experienced when brought back to the surface too quickly. Some of these attempts proved fatal. The flight of the *Zénith* from Paris in 1875 resulted in the deaths of two balloonists. Only the more experienced and probably better-acclimatized balloonist Gaston Tissandier survived to describe the experience.

Left: Illustration showing the 1875 flight of the *Zénith* with pilot Théodore Sivel, engineer Joseph Crocé-Spinelli, and civilian Gaston Tissandier aboard. In spite of precautions that included carrying bags containing a mixture of oxygen and air, Sivel and Crocé-Spinelli both died of asphyxiation when they reached an extremely high altitude. Inset: U.S. Air Force Capt. Joseph W. Kittinger makes his record-breaking skydive from over 101,000 feet (30,784 m).

Pressurized Suits and Plastics

Wiley Post in the high-altitude pressure suit he designed.

Despite the risks, scientists continued to send manned high-altitude balloons into the upper atmosphere. The final such flight of the nineteenth century occurred on December 4, 1894.

In the *Phoenix*, Arthur Berson, a scientist with the Prussian Meteorological Institute, reached 30,000 feet (9,144 m) above Strasburg, Austria, where he recorded the temperatures he found. In 1901 Berson and another scientist with the institute, Reinard Süring, ascended in the *Preussen* to a height of 35,500 feet (10,820 m), a record that stood until broken in 1931 by Swiss balloonist Auguste Piccard.

While some adventurers were still willing to pilot research balloons, scientists chose to develop unpiloted meteorological research balloons, a method that was both cheaper and safer than using manned aircraft. The balloons carried lightweight precision instruments and approached altitudes of 50,000 feet (15,240 m). One such experiment led to the discovery of the stratosphere by meteorologist Léon Teisserenc de Bort in 1899.

Post Goes to New Heights

Among his many accomplishments, pilot Wiley Post helped to develop the first practical pressure suit, which led to pioneering high-altitude flight. In a bid to win the MacRobertson Race, a long-distance trek from England to Australia, Post believed the key to winning would be to fly in the sub-stratosphere—between 30,000 and

L'ENTREPREMANT

A flight from Hamburg, Germany, by Belgian optician and stage illusionist Etienne Robertson and a music teacher named Lloest in July 1803 was among the first made expressly for scientific purposes. Ascending in *L'Entrepremant*, a balloon that had served Napoleon in 1794, the balloonists claimed to have climbed to a record 23,526 feet (7,171 m). This claim has since been disputed because of Robertson's outlandish description of the effects of the high altitude and because it was doubtful that two people could have ascended to such an altitude with a balloon as small as *L'Entrepremant*.

40,000 feet (9,144 and 12,192 m)—where the air is thinner and a plane can travel faster. But since the thin atmosphere makes it impossible to breathe at such altitudes and Post's Lockheed Vega, the *Winnie Mae,* was neither airtight nor pressurized, he set out to develop a pressure suit that would allow him to breathe as if he were at 5,500 feet (1,676 m). With the assistance of the B. F. Goodrich Rubber Company, Post experimented. After rejecting two early models, he successfully tested a suit on September 5, 1934, during a flight over Chicago at 40,000 feet.

By the time Post had perfected his pressure suit the MacRobertson Race was over. He then decided to use it in his effort to set a new transcontinental

flight record. By traveling 2,035 miles (3,275 km) in 7 hours and 19 minutes he succeeded in establishing a new airspeed record during one of these attempts. The *Winnie Mae* reached 340 miles per hour (547 km/h) during that flight, more than a third faster than its normal maximum airspeed, and Post proved that high-altitude flight was the key to faster airspeed.

Jean Piccard and the Plastic Balloon

In 1936 Jean Piccard, brother of Auguste Piccard, developed and launched the first plastic film balloon, the forerunner of modern high-altitude research balloons. Jean also devised a concept for using multiple balloons and in 1937 made the first such manned ascent, climbing in the *Pleiades* to 11,000 feet (3,353 m) using a cluster of 92 balloons attached to a metal gondola.

After World War II Jean Piccard returned to his work on plastic balloons. He worked with engineer Otto C. Winzen to find a suitable material. They finally settled on polyethylene and then worked on how to manufacture balloons from sheets of this plastic that were only a thousandth-of-an-inch thick. This material allowed for extremely lightweight balloons with a marked increase in lifting ability.

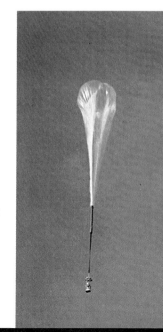

A Project Skyhook balloon shortly after its release from the deck of an aircraft carrier.

Toward the Frontier of Space

After World War II, the first large-balloon launch was organized by Piccard and Winzen on September 25, 1947. Four launches were planned in an operation called Project Skyhook. The polyethylene balloons with a capacity of 100,000 cubic feet (2,830 cubic meters) carried only 70 pounds (32 kg) of equipment. These experiments were so successful that the U.S. Navy abandoned the idea of human balloon flights and focused solely on unmanned research. Thousands of successful Skyhook flights, made over the following decades, added immeasurably to the understanding of the Earth's atmosphere.

Reaching New Heights

Based on the success of Project Skyhook, in 1954 the U.S. Office of Naval Research made plans to entrust human lives to a thin film of poly-ethylene plastic. Five Strato-lab flights were made over the next six years. Lieutenant commanders Malcolm D. Ross and M.L. Lewis reached a record altitude of 76,000 feet (23,165 m) in 1956 in *Strato-lab I*. As flights progressed, altitudes increased, and Ross and Lt. Cdr. Victor A. Prather ascended to 113,739 feet (34,668 m) in *Strato-lab V* on May 4, 1961.

The Strato-lab experiments made a number of important contributions to

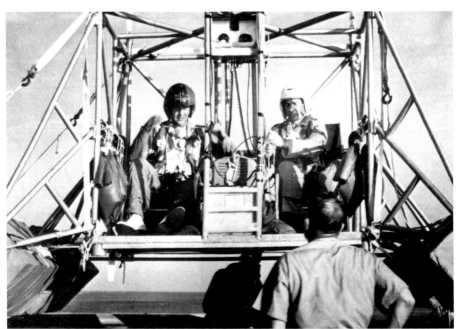

Lt. Cdr. Victor Prather, left, and Cdr. Malcolm Ross landing after a test flight.

"I will ascend above the heights of the clouds; I will be like the most High."

—Isaiah 14:14

These ascents were made as part of a space medicine research program meant to determine if humans were physically and psychologically capable of extended travel at extreme altitudes. A highlight of the program was the parachute bailouts by Kittinger from altitudes in excess of 101,000 feet (30,784 m). Prior to Kittinger's attempts, it was not known whether humans could survive a jump from the edge of space.

Capt. Joseph Kittinger Jr. waits in the balloon gondola (right) as the balloon is filled with helium for the *Excelsior I* test jump in New Mexico. Kittinger set the record for a freefall parachute jump at 76,400 feet.

the human spaceflight program as well as to astronomy and meteorology.

Project Manhigh

During the same period, the U.S. Air Force began a project to study high-altitude escape procedures. Project Manhigh consisted of three balloon flights to the edge of space. On the first ascent in June 1957, Capt. Joseph W. Kittinger II reached an altitude of 77,099 feet (23,499 m). In that same year, Maj. David Simons reached 101,377 feet (30,899 m) and Lt. Clifton McClure achieved 98,097 feet (29,899 m). Each Manhigh flight carried a single pilot in a small phone booth–like capsule crammed with instruments.

ATMOSPHERIC PRESSURE

Atmospheric pressure drops as an aircraft rises. This can cause a wide variety of problems. Not only do humans need oxygen to breathe, but also atmospheric pressure behind it to help force it into the lungs. The lower the pressure, the harder it is to breathe, no matter how much oxygen the air may contain. Lower air pressure also lowers the boiling point of water and other liquids. For each thousand feet above sea level, the boiling point of water drops almost 2°F (0.5°C). This can have a catastrophic effect on the human body.

CHAPTER 11

THE QUEST FOR SPEED

ULTIMATE SPEED HAS BEEN a sought-after goal since the invention of the airplane. Air races have been held almost from the very start, driving rapid development of new technologies that were eventually applied to conventional aircraft.

The Schneider Trophy seaplane races of the 1930s facilitated the development of the British Supermarine racers. These planes were designed by Reginald Mitchell, who went on to develop the Spitfire fighter of World War II fame. The Supermarines were powered by Rolls-Royce engines, and that company developed the Merlin engine for Mitchell's fighter plane. Without the Spitfire and the Merlin, England's chances of winning the Battle of Britain would have been significantly lessened.

In the United States, the bizarre but lightning-fast Gee-Bee Super Sportster set land-plane speed records, and Howard Hughes worked on his design for a state-of-the-art plane that eventually flew more than 100 miles an hour (320 km/h) faster than any military plane in the world.

Left: NASA's X-43A Pegasus. This aircraft set the world record for jet-powered speed in 2004 at Mach 9.6. Inset: Howard Hughes with his Hughes Racer.

Pushing the Envelope

German engineers Ernst Heinkel (left) and Hans Von Ohain, who developed the first jet-propelled aircraft.

The 1934 MacRobertson Air Race covered an 11,300-mile (18,186 km) course from Mildenhall, near London, to Melbourne, Australia. De Havilland Aircraft was determined to enter the fastest plane the contest had ever seen. The result was the DH-88 Comet, a twin-engine plane as streamlined as a modern jet. It won the race in a blazing 71 hours. The bright red Comet demonstrated what could be done when streamlined design was coupled with relatively small, efficient, air-cooled, supercharged in-line engines. The Comet eventually evolved into the De Havilland Mosquito, a fighter and bomber that would outrace many other fighters during World War II.

Perhaps more important than the Comet's victory were the planes in second and third place. A Douglas DC-2 and a Boeing 247D both came in ahead of other DH-88s. What makes this significant is the fact that these two planes were not specially built for racing, as the Comet had been, but were purely commercial, passenger aircraft.

By the end of the decade, two aircraft had been developed that were to profoundly affect the quest for speed. One was the rocket-propelled He-176 and the other the turbojet-propelled He-178.

Through the Barrier

The *He* in He-176 and He-178 stands for Heinkel. Professor Ernst Heinkel (1888–1958) was a German aeronautical engineer who designed some of the most advanced and successful aircraft of the early twentieth century.

In 1935 Heinkel met rocket scientist Wernher von Braun (1912–77). Heinkel provided the research facilities where von Braun produced the rocket engine for Heinkel's He-176. It made its first flight in 1939, achieving a speed of more than 500 miles per hour (804 km/h).

Meanwhile, Heinkel hired engineer Hans Joachim Pabst von Ohain (1911–98), who had been developing gas turbine engines. The first successful engine test was run in September 1937, and by August 1939 the He-178 had taken to the air—the first jet-propelled aircraft in history. The little monoplane, measuring just 24 feet (7.3 m)

SPECIFICATIONS

HE-176
Length: 17 ft 1 in (5.2 m)
Wingspan: 16 ft 5 in (5 m)
Height: 4 ft 11 in (1.5 m)
Empty weight: 1,980 lbs (900 kg)
Engine: HWK RI-203
Maximum speed, theoretical: 470 mph (750 km/h)
Maximum speed, attained: 215 mph (345 km/h)

"The important thing in aeroplanes is that they shall be speedy."

— BARON MANFRED VON RICHTHOFEN

long, had a top speed of 435 miles per hour (700 km/h). By the end of World War II, jet aircraft were easily attaining speeds well over 500 miles per hour (805 km/h).

The Sound Barrier

As propeller-driven and jet aircraft both approach the speed of sound (760 mph/1,223 km/h at sea level), strange things begin to happen. Aeronautical engineers discovered that at such speeds, the airflow around an aircraft either created destructive shock waves or such high drag that the plane simply could not go any faster—hence the term *sound barrier*. Pilots had begun noticing these effects as early as World War II, when planes like the Spitfire went into power dives that took them to nearly 90 percent of the speed of sound. Once engineers realized that the

jet and rocket aircraft of the near future would easily achieve the power necessary for trans-sonic speeds, they began to consider the design requirements of such an aircraft: one that would be able to slip harmlessly through the sound barrier.

The Americans were the first to achieve success in breaking through, with the rocket-powered Bell X-1. Pilot Charles "Chuck" Yeager (b. 1923) broke the sound barrier on October 14, 1947, with a top speed of Mach 1.015, which was 670 miles per hour (1,078 km/h) at the altitude he was flying. The X-1 was followed up by the rocket-propelled X-2, specially designed to deal with the tremendous heat created by air friction at high speeds. Engineers had replaced the X-1's aluminum skin with stainless steel for the X-2, which achieved a speed of Mach 3.2 (2,094 mph/3,370 km/h) in 1956.

SPECIFICATIONS

DE HAVILLAND 88 COMET
Wingspan: 44 ft (14.41 m)
Length: 29 ft (8.84 m)
Height: 10 ft (3.05 m)
Engine: Two 230-hp De Havilland Gypsy Six R inline piston engines
Maximum speed: 237 mph (381 km/h)

Above: Chuck Yeager with the Bell X-1 nicknamed Glamorous Glennis.

Left: NASA's Robert T. Jones, the first American scientist to identify the importance of swept-back wings in achieving supersonic flight.

Supersonic Flight Goes International

An SR-71 (bottom) taking off with an F-18 in the background.

The British aircraft Electric P.1A exceeded the speed of sound in level flight in August 1954, paving the way for the Lightning, which was the Royal Air Force's first supersonic fighter. The first jet to break the sound barrier was the British DH-108, a bat-winged aircraft developed at the end of World War II. It made its first flight in 1946 and broke the sound barrier in September of 1948. A French pilot flying a Sud-Ouest SO-9000 Trident broke the sound barrier in March of 1953. The Trident was a hybrid, with two wing-tip-mounted turbojets and tail-mounted rockets that boosted it to Mach 1.6.

The MiG-25 was the fastest fighter plane to ever enter service. Developed in the Soviet Union during the 1960s, flight testing on the MiG-25 began in 1964. The similar MiG-31 followed in the 1970s. Before long all existing speed, altitude, and time-to-climb records were broken by the Ye-266, a special stripped-down, high-performance aircraft.

The Fastest

The fastest jet aircraft in the world was the SR-71 Blackbird. It achieved Mach 3.35 (2,275 mph/3,661 km/h) at 80,000 feet (24,384 m), a world record. The development of a Mach 3 high-altitude interceptor in the late 1950s was the beginning of the Blackbird

HOW ROCKETS WORK

In the late 1600s English scientist and mathematician Sir Isaac Newton (1643–1727) developed laws that explained how and why things move the way they do. Although he never mentioned the rocket per se, the third of his three laws of motion described how rockets work. It states that for every action there is an equal and opposite reaction. When a gun is fired, the burning gunpowder forces a bullet from the barrel. At the same time, the gun moves in the opposite direction. This opposite movement is called recoil. If a gun shoots out a continuous stream of bullets, as a machine gun does, the recoil is continuous. This is what happens in rocket propulsion. The gas molecules produced by the burning fuel act like trillions of tiny, individual bullets. As each one is ejected from the rear of the rocket, the rocket moves in the opposite direction, just as Newton's law projected it would.

design. Uneasy about the vulnerability of its U-2 spy planes, the Central Intelligence Agency realized that it needed a high-speed, high-altitude replacement. This led in 1958 to the development of the SR-71, created to achieve a maximum speed of over Mach 3 and a ceiling of up to 85,000 feet (25,908 m), which required new aeronautical technology and construction methods.

Although it provided invaluable reconnaissance capabilities during the Cold War, maintenance and operation of the SR-71 proved to be very expensive. The U.S. Air Force, bowing to budget restraints and reduced funding, was forced to retire the aircraft in 1990. Only three models of the SR-71 remain in service, operated by NASA as part of its high-speed research program.

SPECIFICATIONS

SR-71
Wingspan: 55 ft 7 in (16.94 m)
Length: 107 ft 5 in (32.74 m)
Height: 18 ft 6 in (5.64 m)
Maximum speed: Mach 3 to 3.5
Engines: Two Pratt & Whitney
J58 turbojets

Congress ordered the reactivation of a portion of the SR-71 fleet in 1994 in the belief that the aircraft would be useful in providing reconnaissance during the Gulf War. NASA loaned the Air Force its three SR-71s since the aircraft that had been in storage were no longer considered airworthy. However, the Air Force again retired the SR-71 in 1998 while NASA continued to use its three research aircraft until 2001, when they were at last retired permanently.

The MiG-31 was nicknamed Foxhound in the United States. This Soviet fighter was fitted with one of the most powerful fighter radars in the world.

Rocket Riders

Scores of inventors had designed rocket-propelled balloons and airships throughout the nineteenth and early twentieth centuries, yet few, if any, got off the drawing board. In the mid-1920s Austrian engineer Max Valier (1893–1930) began to experiment with rocket-powered cars and other vehicles. He persuaded automobile manufacturer Fritz von Opel (1899–1971) to sponsor the construction and flight of a rocket-powered airplane. They selected an advanced type of glider, named the *Ente* (German for *duck*), and adapted it to carry rockets in the rear of its fuselage. Tests were carried out by the research institute of the Rhine-Rossitten Experimental Association, which provided the glider and the test models. On June 11, 1928, after several days of testing, pilot Friedrich Stamer (1897–1969) became the first man to fly

A German rocket car with Fritz von Opel in the driver's seat and Max Valier holding the car with his right hand.

in a rocket-powered aircraft. The total length of the flight was between 4,200 and 4,920 feet (1,280 and 1,308 m) and lasted between 40 and 80 seconds.

Commenting on the flight, *Scientific American* magazine observed that, "On the whole we are inclined to think that the rocket as applied to the airplane might be a means of securing stupendous speeds for a short interval of time, rather than a method of very speedy sustained flight."

Rockets with Wings

Designing, building, and flying rocket-powered aircraft became something of a minor trend during the 1930s. Although most of these experiments were carried out by enthusiastic amateurs, others had begun to take the idea of rocket-powered aircraft seriously. Where most previous rocket-propelled aircraft had been little more than conventional gliders with rockets fitted to them, aircraft powered by rocket propulsion were being designed from scratch.

Between 1931 and 1933, Russian engineer Fridrikh A. Tsander (1887–1933) and his fellow members of GIRD (Group for the Study of Reaction Motion) began developing a rocket-propelled aircraft. A GL-1 glider would be powered by Tsander's liquid-fueled ER-2 engine. The work was supervised by Sergei P. Korolev (1906–66), who would

"Hey Ridley, that Machometer is acting screwy. It just went off the scale on me."

—GENERAL CHARLES "CHUCK" YEAGER ANNOUNCING THE FIRST TIME HE WENT SUPERSONIC

go on to mastermind the Soviet space program.

Unlike Valier's solid-fuel rockets, Tsander's engine could be throttled. The glider was a triangular flying wing made of wood with a wingspan of 39.7 feet (12.1 m) and a length of 10 feet (3.09 m). Although the engine received final approval on December 23, 1932, and tests were begun, the project was eventually abandoned so that attention could be directed toward developing the RP-218.

The RP-218 rocket glider was designed by Korolev in 1933. The glider was essentially a modified SK-9 two-seat airframe fitted with an ORM-65 rocket. On February 28, 1940, a Soviet P-5 military airplane towed the newly designated RP-318-1 into the air. Released at an altitude of 8,500 feet (2,600 m), the plane increased its speed from 50 to 87 miles per hour (80 to 140 km/h) in just 5 to 6 seconds, and within 110 seconds its altitude had gone up by some 984 feet (300 m).

Both the Soviet Union and Germany actively pursued the development of rocket-propelled aircraft, such as the

Soviet rocket scientist Sergei Korolev.

SPECIFICATIONS

RP-318-1
Length: 24 ft 5 in (7.44 m)
Wingspan: 55 ft (17 m)
Weight: 1,540 lbs (700 kg)
Engine: RDA-1

Russian rocket fighter BI-1, during World War II. The BI-1 evolved into a series of rocket-powered Soviet aircraft between 1941 and 1947, from the Maljutka, developed and flown during the war, to the postwar Mikoyan I-270.

THE FIRST AMERICAN ROCKET PLANE

On June 4, 1932, stunt pilot William G. Swann flew a rocket-propelled glider from Bader Field at Atlantic City, New Jersey. The first American to fly a rocket-propelled aircraft, he reached an altitude of 200 feet (61 m) in his Steel Pier Rocket Plane, a craft powered by 12 solid-fuel rockets. Swan rode below the wing of the converted high-wing monoplane glider in an open seat.

The Komet

Nazi Germany produced several so-called *Wunderwaffen* (wonder weapons) during World War II. The Messerschmitt Me-163Vl Komet rocket plane was one such wonder. On October 2, 1941, German pilot Heini Dittmar (1911–60) made the first powered flight of the Komet. Once the plane was towed to an altitude of 13,000 feet (3,962 m), Dittmar took off—reaching a speed of 623.85 miles per hour (1,004 km/h). The Komet, the first aircraft to exceed the speed of 1,000 kilometers per hour, experienced violent turbulence at Mach 0.84. In another test flight, the Me-163, piloted by Hanna Reitsch (1912–79) reached a speed of 624 miles per hour (1,004 km/h), with Reitsch becoming the first woman ever to fly a rocket-powered aircraft.

The Luftwaffe planned to develop an interceptor version of the Komet, which resulted in the Me-163B, utilizing the new Walter HWK 509 rocket

motor. The plane's first success was achieved on August 5, 1944, when three Me-163Bs shot down three American P-51D Mustang fighters.

Although rocket-powered aircraft such as the Messerschmitt Me-163 and the Japanese Ohka are among the best-known rocket planes of World War II, other nations also developed rocket aircraft, though few if any saw active service. These included such aircraft

KAMIKAZE ROCKETS

The Japanese developed their infamous Ohka rocket plane late in the course of World War II in an effort to improve the results of the Kamikaze suicide bombers, in which a pilot would deliberately dive his bomb-laden aircraft into an enemy ship. Carried by a bomber to within about a dozen miles of an enemy aircraft carrier, the Ohka would be released, using its three rocket motors to propel it at a speed—over 600 miles per hour (966 km/h)—beyond what was available to American fighters. In the nose of the rocket plane was more than a ton of high explosives, which would be detonated when the pilot dived into the ship.

as the Soviet 216-1, Kostokov 302, and Polikarpov Malyutka interceptor, the British DH-100 Swallow, and the never-built Japanese J8MI Shusui.

In the United States, Northrop produced a feasibility study for a rocket-powered interceptor, which was followed by an Army contract in 1942. This led to the development of the XP-79 all-magnesium rocket fighter-interceptor. Its speed was designed to be in excess of 500 miles per hour (805 km/h), powered by an Aerojet rocket motor with 2,000 pounds (907 kg) of thrust. The final version was designated the MX-324. The pilot flew the small aircraft lying on his stomach, which allowed for a slimmer wing cross-section and also enabled the pilot to withstand greater g-forces.

The MX-324 made its first powered flight on July 5, 1944, propelled by an Aerojet engine with 200 pounds (91 kg) of thrust. It was the first liquid-fuel rocket-powered aircraft to fly in the United States.

THE SECOND AMERICAN ROCKET PLANE

In August 1941 Lt. Homer A. Boushey (1909–2001) became the second American to fly a rocket-powered aircraft. The plane was a commercially manufactured Ercoupe, a small light plane. Under the auspices of the California Institute of Technology and Aerojet-General founder Theodore von Karman (1881–1963), the experiment was performed for the U.S. Army Air Forces. The plane was fitted with six standard solid-fuel JATO (jet assisted takeoff) rockets. On the first flight, the plane took off from March Field in Riverside, California, using its rockets only as a booster. Later, Boushey made several flights with the plane's piston engine turned off, taking off and flying by rocket power alone.

SPECIFICATIONS

ME-163 KOMET
Wingspan: 30 ft 7 in (9.32 m)
Length: 19 ft 2 in (5.84 m)
Height: 9 ft 1 in (2.77 m)
Empty weight: 4,200 lbs (1,905 kg)
Maximum speed: 596 mph (960 km/h)
Engine: One Walter HWK 509A-2 rocket
  motor

The X-Planes

The Ryan X 13 Vertijet, an experimental vertical takeoff and landing (VTOL) aircraft that was tested in the 1950s.

If any series of aircraft defines the moniker extreme, it may be the series of experimental, or "X," aircraft that have been developed since 1950. Among these are some of the most unusual aircraft ever conceived. They include many important research aircraft, such as the X-15. And while some X-planes are almost indistinguishable from conventional aircraft, others are just short of bizarre.

Some experimental aircraft were designed to study the effects of flying at high speeds and altitudes, while others were meant to study aerodynamics, new propulsion systems, or aircraft configurations. Several laid the groundwork for the next generation of spacecraft. The Martin Marietta X-24A and X-24B, two of a group of aircraft called lifting bodies, helped to develop radical new reentry technologies for present-day and future spacecraft.

Straight Up and Down

A number of experimental aircraft were developed in the long search for a vehicle that could take off and land vertically and fly like a conventional aircraft once airborne. The Ryan X-13 Vertijet loosely resembled the conventional delta-winged jet fighters of the 1950s—except that it could take off and land on its tail. It made its first conventional takeoff and flight

on December 10, 1955, and a vertical takeoff and landing the following May. A year later, in April 1957, it took off vertically, turned to level flight, and then landed vertically. This was the first time any aircraft had achieved this, proving the viability of vertical takeoff and landing (VTOL) aircraft.

The Bell X-14 and the Hiller X-18 tried to accomplish this same feat in two entirely different ways. The X-14 was built to study the feasibility of vectoring the exhaust of the plane's jet engine in order to lift the plane vertically from the ground. The X-14 made

"I flipped on the data switch . . . Usually, at about this time, I began to wish very much that I had taken Mother's advice and had actually attended dental school."

—WILLIAM BRIDGEMAN, DESCRIBING A TEST FLIGHT OF THE SKYROCKET

its first successful test flight in 1954. This technology was later perfected for the famed Harrier fighter. The X-14 was employed for nearly 25 years, until its retirement in 1987, as a training vehicle for VTOL pilots.

The Hiller X-18 took an entirely different approach to the VTOL problem. Outwardly resembling a large, conventional twin-engine aircraft, it could swing its entire wing—engines and all—to a vertical position. Making its first flight in 1959, the X-18 was the first major project to investigate the use of tilt-wing, short/vertical takeoff or landing (S/VTOL) aircraft. The Vought XC-142 and the controversial V-22 Osprey are direct descendants of the X-18. The Curtiss-Wright X-19 took a similar tack, but instead of swinging the entire wing to a vertical position, only the engine pods were pivoted for vertical flight.

Perhaps the oddest-looking of all the experimental VTOL craft was the Bell X-22a, which was developed to test the viability of ducted lift propellers. For all its strange appearance, the aircraft was very successful, making test flights from 1966 to 1984.

Strange Wings

The Grumman X-29 was developed to determine the effectiveness of forward-swept wings, as well as the use of exotic materials in airframe construction. The 35-degree forward sweep of the wings promised improved agility and proof against spinning and stalling, as well as good handling characteristics at low speeds. The first prototype of the plane flew in 1984, making 242 flights before being grounded in 1988. The second, X-29A, first flew in 1989, adding another 132 flights. No other X-plane has completed as many flights as these two together.

The McDonnell Douglas X-36 was a remote-controlled quarter-scale model of a proposed tailless fighter aircraft. It completed 31 flights in 1996 and 1997. The extreme maneuverability of the aircraft and its agility and stability at high and low speeds suggest that the design may have great potential as a combat aircraft.

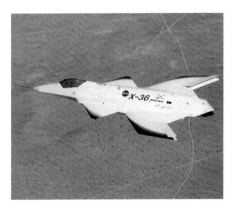

An X-36 tailless fighter research aircraft.

SPECIFICATIONS

GRUMMAN X-29A
Wingspan: 27 ft 2 in (8.29 m)
Length: 53 ft 11 in (16.44 m)
Height: 14 ft 3 in (4.36 m)
Maximum level speed: Mach 1.5
Engine: One General Electric F404-GE-400
   turbofan

Above: The X-243 was the last aircraft of NASA's manned lifting-body program.

Below: An X-38 lifting body aircraft in free flight.

Several X-planes have been developed with space travel in mind, such as the USAF/NASA Martin Marietta X-24A and X-24B. More recently, the Scaled Composites X-38 was developed in the late 1990s to study a possible design for a crew return vehicle (CRV) that could be used as an emergency lifeboat of sorts for the crews of the International Space Station. Based on earlier lifting-body technology, the unpiloted X-38 was air-dropped, deploying a paraglider for landing.

The Lockheed X-33 was the model for a manned, single-stage, multiple-use, heavy-lift spacecraft with the name VentureStar. Only a one-third scale version of the spacecraft was created before the project was canceled by NASA in 2001.

NASA's X-43A Hyper-X is already the fastest aircraft ever flown. The unpiloted, scramjet-powered Hyper-X is laying the foundation not only for future hypersonic transports but recoverable boosters for spacecraft as well.

The Skyrocket

On January 26, 1951, a rocket-powered version of the Douglas D-558-II No. 2 made its first flight. Pilot William Bridgeman (1916–68) took the air-launched Skyrocket from 32,000 to 41,000 feet (9,754 to 12,487 m), reaching a speed of Mach 1.28 on a level run.

On November 20, 1953, the D-558-II No. 2 Skyrocket, piloted by A. Scott Crossfield (1921–2006), became the first aircraft to fly at twice the speed of sound (Mach 2, or 1,291 mph/2,078 km/h). The sleek, white rocket was designed and built by Douglas Aircraft and powered by a Reaction Motors four-chambered rocket engine virtually identical to the one used in the Bell X-1.

SPECIFICATIONS

X-43A HYPER-X
Wingspan: 5 ft (1.5 m)
Length: 12 ft (3.7 m)
Weight: 3,000 lbs (1,300 kg)

As with the Bell X-1, safety considerations eventually caused the Skyrocket to be modified for air launching. It was carried aloft slung below the bomb bay of a specially adapted B-29 Superfortress. As it was able to carry additional rocket fuel in the space given up by its jet engine, the D-558-II was able to exceed Mach 2.

SPECIFICATIONS

DOUGLAS D-558-II SKYROCKET
Length: 42 ft (12.8 m)
Wingspan: 25 ft (7.6 m)
Height: 22 ft 8 in (3.8 m)
Empty weight: 9,421 lbs (4,273 kg)
Engine: One Westinghouse J34-WE-40; one
  Reaction Motors XLR-8-RM-5 rocket engine
Maximum speed: 1,291 mph (2,078 km/h)

CHAPTER 12

FLYING INTO SPACE

IN 1950, DAVID G. STONE, head of the pilotless aircraft research division of the National Advisory Committee for Aeronautics (NACA) proposed that a large supersonic airplane could launch a smaller manned second stage at Mach 3 that would then accelerate to hypersonic speeds. He suggested that the second stage should be a modified Bell X-2, equipped with reaction controls and two solid fuel boosters that could be jettisoned. The boosted X-2 would be able to attain Mach 4.5 and orbital altitudes. Stone's plan was evaluated by a NACA study group that ultimately rejected it on the grounds that the X-2 was too small to use as a hypersonic research aircraft. They decided that an entirely new vehicle was required, and ten years later the X-15 made its first powered flight.

The development of the X-15 research program began in 1954 as a joint venture sponsored by NACA, the U.S. Air Force, the U.S. Navy, and private industry. The purpose of the aircraft was to study the problems of hypersonic and space flight at speeds exceeding Mach 6.6 and altitudes of 12 to 50 miles (19 to 80 km). North American Aviation, Inc. was selected to design three X-15 research planes and Reaction Motors, Inc. won the contract for development of the XLR99 rocket engine.

Left: Artist's rendering of the *X-33* in flight. Lockheed Martin developed the X-33, also known as VentureStar, as a single-stage to orbit, reusable launch vehicle for NASA, but the program was canceled in 2001. Inset: *SpaceShipOne* and the X Prize team members.

The X-15 and the Dyna Soar

The aircraft that was ultimately developed proved to be the link between atmospheric flight and spaceflight. It had wings and control surfaces, so it could fly within the atmosphere, but it was also equipped with wingtip and nose thrusters so that it could maneuver in the near-vacuum at the fringes of outer space. The X-15 was so much like a true spaceship that its pilots were required to wear

Three X-15 aerospace planes were built. One was destroyed, but the remaining two had a long and successful service life, accumulating a total of 199 flights. Capable of speeds in excess of six times the speed of sound—speeds at which air friction heated its nickel-alloy skin to a temperature of 1,200 degrees F (439°C)—the X-15 was one of the most successful research aircraft ever created.

A Dyna Soar in Space

The United States nearly had a space shuttle in the 1960s. The X-20 Dyna Soar (Dynamic Soarer) was a single-pilot, manned, reusable space plane developed through an Air Force program that ran from October 1957 to December 1963. Intended for use by the military, the X-20's missions would have included reconnaissance, bombing, space rescue, satellite maintenance, and the destruction of enemy satellites.

Above: Neil Armstrong stands before an X-15 after a research flight.

Above right: Illustration depicting the launching of a Dyna Soar aircraft.

full-pressure space suits when flying it, and the USAF pilots who eventually took it above 50 miles (80 km) were awarded astronaut's wings. The maximum altitude reached by an X-15 was over 67 miles (108 km), which was not exceeded until the flights of the *Mercury* spacecraft.

Instead of a ballistic capsule, Dyna Soar would have been a small lifting body that would glide to Earth under the control of the pilot and land at a preselected site like a conventional aircraft.

Dyna Soar had its roots in the Silver Bird rocket bomber invented by Eugen Sänger (1905–64) during World War II. Following the war, many German scientists were taken to the United States as part of an endeavor code-named Operation Paperclip. Among the group of scientists was Dr. Walter Dornberger (1895–1980), the former head of Germany's wartime rocket program, who had detailed knowledge of the Sänger project. Working for Bell Aircraft during the 1950s, he and aerospace engineer Krafft Ehricke (1917–84) tried to interest the U.S. military in a similar system. This resulted in the establishment of the Dyna Soar program, to develop an orbital weapons system and a successor to the X-15 research program. The space plane would be launched with an Air Force Titan III booster.

Seven Dyna Soar astronauts were selected in 1960, and at the end of 1962, after a series of successful engine tests, the USAF officially unveiled the hitherto secret X-20. Ten flights were planned between January 1966 and March 1968, beginning with unmanned flights and ending with manned Earth orbital missions.

Unfortunately, disputes about the project's funding and costs, choice of booster, and a lack of clear focus regarding the project's goals—was it to be a research vehicle, a weapons system, or a combination of both?—eventually caused the X-20 Dyna Soar program to be canceled in 1963.

SPECIFICATIONS

X-20 DYNA SOAR
Length: 35 ft 4 in (10.77 m)
Wingspan: 20 ft 10 in (6.34 m)
Height: 8 ft 6 in (2.59 m)
Empty weight: 10,395 lbs (4,715 kg)
Engine: One Martin Trans-stage rocket engine
Maximum speed (proposed): 17,500 mph (28,165 km/h)
Range (proposed): Earth orbit at 22,000 nautical miles (40,700 km)
Service ceiling (proposed): 530,000 ft (160,000 m)

SPECIFICATIONS

X-15
Length: 50 ft 8 in (15.45 m)
Wingspan: 22 ft 4 in (6.8 m)
Height: 13 ft 6 in (4.12 m)
Empty weight: 14,600 lbs (6,620 kg)
Engine: One Thiokol XLR99-RM-2 liquid-fuel rocket engine
Maximum speed: 4,520 mph (Mach 6.85, 7,274 km/h)
Range: 280 miles (450 km)

The Space Shuttle

The Space Shuttle, officially known as the Space Transportation System, or STS, is the direct descendant of the earliest rocket-propelled aircraft. Rockets that were specially constructed to ferry personnel and materials to and from orbit had long been part of the space stations planned from the 1920s on.

The Shuttle was also a product of lifting-body research. Lifting bodies are aircraft that depend on the shape of the fuselage instead of wings for lift. They are strange-looking things that resemble flying irons. However, their shape and flying qualities make them ideal models for spacecraft that need to glide to a landing after re-entry. For instance, their flat bottoms make perfect heat shields and their lack of wings reduces air friction.

President Richard Nixon created a Space Shuttle task group in 1969. Its first job was to decide what the best configuration for a shuttle might be. The committee eventually decided that it should consist of two rocket-powered aircraft, very much like the shuttle proposed by Dornberger and Ehricke 20 years earlier. One of the rockets would be a booster, the other an orbiter. The two would take off together. At a certain altitude the two would separate. The booster would be flown back to its base, while the orbiter would continue on into orbit. It was to be a fully reusable aircraft, which meant that there would be no expensive throwaway stages like those in the other models.

An Expensive Endeavor

Budget considerations ate away at the Shuttle design. While it may have been cheaper in the long run to have a reusable booster, initial development costs limited the project to the orbiter alone. It would have to be launched using expendable components.

More money could be saved, it was decided, if the orbiter did not have to carry its own fuel, which could be contained in an expendable external tank. This way, the Shuttle could be made smaller and less expensive. Strap-on solid-fuel booster rockets would have to be added, however, since the Shuttle's main engines would not be able to provide enough thrust on their own to get the spacecraft into orbit.

By 1972 the Shuttle had more or less reached its final configuration—a lifting-body orbiter with an external fuel tank and two solid-fuel boosters. The relatively small orbiter was dwarfed by these components. While the space plane was only 122 feet (37 m) long, the external tank reached a length of 154 feet (47 m), and the two boosters were each 149 feet (45 m) tall.

The first orbiter, named *Enterprise* at the insistence of legions of Star Trek fans, was first tested in 1977. These were unpowered glide-and-landing tests, with the Shuttle launched from the back of a specially adapted Boeing 747. As big and clumsy-looking as the Shuttle appeared, its performance surprised its test pilots, who compared its handling characteristics to a fighter jet.

Space Shuttle in Detail

The Space Shuttle, a reusable launch vehicle, consists of three different main assemblies. First is the reusable orbiter. This is a delta-winged aircraft with a single vertical stabilizer. It is equipped with ordinary control surfaces for flying within the atmosphere and thrusters for maneuvering in airless space. The cabin consists of three

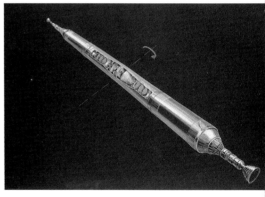

Top: Space Shuttle prototype *Enterprise* as it is launched by a 747 Shuttle carrier aircraft (1977).

Bottom: President Richard Nixon created the Space Task Group to study the possibilities of manned space flight. This illustration shows one idea considered by the group—a spacecraft for a manned mission to Mars.

Opposite page: Space Shuttle *Atlantis* lifts off to a rendezvous with the International Space Station.

Astronaut Michael L. Gernhardt holds on to the end of a Canadarm on the Space Shuttle *Atlantis* during a mission to complete the second phase of construction on the International Space Station.

of excess heat. For this reason, the doors are kept open while in orbit. Inside the bay is the remote manipulating system—a remote-controlled mechanical arm called the Canadarm, because it is manufactured by a Canadian company. It allows the astronauts on board the Shuttle to move objects in the bay without having to leave the cabin.

The second part of the STS is the huge external tank. It contains the liquid hydrogen and liquid oxygen that fuel the three Shuttle main engines—a total of 535,000 gallons. The Shuttle itself contains no fuel (except for the small amounts of propellant needed to work its thrusters).

Finally, there are the two solid-fuel boosters (officially Solid Rocket Boosters, or SRBs). These 149.6-foot (46 m) tubes are each 12.17 feet (3.8 m) in diameter. Filled with a solid rocket fuel that, once ignited, cannot be stopped, the SRBs are the direct descendants of the skyrockets made nearly a thousand years ago by the Chinese. These boosters provide more than three-quarters of the Shuttle's thrust at takeoff.

levels: the flight deck, the mid-deck, and a utility area. The commander and pilot sit on the flight deck, with two mission specialists sitting behind them. The mid-deck has three more seats as well as a galley, toilets, sleeping facilities, lockers, a hatch, and an airlock.

Behind the cabin is a large payload bay that takes up most of the fuselage—60 feet by 15 feet (18 by 4.5 m), large enough to carry a city bus. The payload doors have heat radiators on their inner surfaces that allow the Shuttle to get rid

Successes and Losses

On April 12, 1981, the Space Shuttle *Columbia* rose from its launch pad at Kennedy Space Center in Florida, carrying Cdr. John Young (b. 1930) and pilot Robert Crippen (b. 1937). After a relatively uneventful mission, *Columbia* landed at Edwards Air Force Base just 54 hours and 20 minutes after takeoff.

As of August 2005 the fleet of five Shuttles had made a total of 114 flights, covering a distance of 430,500,333 miles (692,823,128 km) in 16,557 orbits. Their crews had spent a total of 2.8 years in orbit. During that time the Shuttles had launched satellites and space probes, repaired or returned satellites to Earth, ferried crews to and from the International Space Station, and done scientific research.

The Space Shuttle program had its critics from the very beginning, especially when the original designs began to suffer from budget cutbacks and uncertainties about what functions the Shuttle would actually serve. It rapidly became a victim of endless compromises and over-complexities. Some of these negative developments, such as the use of external fuel tanks and strap-on solid fuel boosters, would prove fatal. When a seal between the sections of one of *Challenger*'s solid-fuel boosters failed, a jet of white-hot gas ignited the contents of the adjacent external fuel tank, causing the entire spacecraft to erupt in a ball of flame only moments after takeoff on January 28, 1986. And on February 1, 2003, the shuttle *Columbia* broke up on reentry when one or more of its thousands of thermal protection system tiles—designed to protect the shuttle from the extreme temperatures encountered during reentry—failed.

NASA is currently studying a number of proposals for second-generation shuttles that will be more specialized in their application and will make use of reusable, liquid-fuel boosters or launch vehicles.

Above: The crew of Space Shuttle *Columbia* after a successful mission.

Below: The *Challenger* tragedy: A bright glow at the top of this image is believed to show the rupture of the spacecraft's liquid oxygen tank.

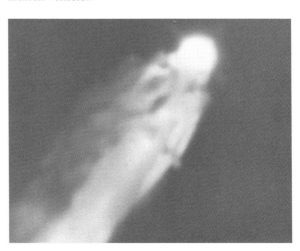

Space Planes

Illustration of the X-30 aerospace plane. This plane was planned to demonstrate technologies necessary for hypersonic cruise vehicles.

A big step toward the achievement of space tourism was the Ansari X Prize, established in 1996 and named for Anousheh and Amir Ansari, who contributed the lion's share of the $10 million prize money. The successful entrant would have to launch a reusable, piloted spacecraft into space (at least 62 miles/100 km) twice within a two-week period. Any government funding was forbidden.

SpaceShipOne

Twenty-six companies and individuals entered the contests. Some of these were fairly conventional designs—for instance, the Canadian Arrow closely resembles the German V-2 of World War II—while other designs were very imaginative.

The prize was finally won by *SpaceShipOne*, a winged space plane that had been created by Burt Rutan's company Scaled Composites. Its two competitive flights were made on September 29, 2004, and October 4, 2004. Funding for the project had

been provided by multi-billionaire space enthusiast Paul Allen. Since *SpaceShipOne*'s victory, some of the other competitors have continued work on their spacecraft. The Canadian Arrow, for instance, has conducted successful tests of its engine and has obtained permission from the Canadian government to create a launch site. So it is possible that *SpaceShipOne* may one day have some competition.

The Quest Continues

A winged vehicle that would be able to take off, fly into space, and then return for a landing like a conventional aircraft has been the goal of aerospace dreamers and engineers since at least the 1920s. Only two orbital space planes have been flown successfully—the United States Space Shuttle and the Soviet Buran. Both were designed to ascend to orbit vertically using conventional expendable launch vehicles, eventually returning to Earth as gliders. Since the spacecraft's wings are not used for the ascent, they are essentially dead weight, limiting the total amount of payload the shuttle can carry. The convenience and economy of a reusable spacecraft, however, was thought to offset this drawback.

SPECIFICATIONS

SPACE SHIP ONE
Length: 165 ft (5 m)
Wingspan: 16 ft (5 m)
Diameter: 5 ft (1.52 m)
Empty weight: 2,645.5 lbs (1,200 kg)
Engine: One N_2O/HTPB SpaceDev Hybrid rocket motor
Maximum speed: Mach 3.09 (3,518 km/h)
Range: 40 miles (65 km)
Service ceiling: 367,453 feet (112,000 m)

"Space isn't remote at all. It's only an hour's drive away if your car could go straight upwards."

—SIR FRED HOYLE

Other space-plane designs have used the vehicle's wings to provide lift for the ascent to space in addition to rocket boosters. The two such spacecraft built to date—the X-15 and *SpaceShipOne*—were launched at high altitudes from a carrier aircraft. The goal of most space-plane engineers is a spacecraft that will be able to take off from a conventional runway, ideally in a single stage, without the need for either expendable or recoverable boosters. There are those who believe that scramjet technology might make such spacecraft possible. Scramjet aircraft burn fuel in air that is moving at supersonic speed. The Rockwell X-30 National Aero-Space Plane (NASP) was an attempt to build a scramjet vehicle, but it failed due to technical issues.

Numerous other nations have proposed their own space-plane programs. The British HOTOL (Horizontal Take-off and Landing) space plane was canceled due to technical and financial issues. One version would have carried 50 passengers halfway around the world in an unpiloted aircraft in less than an hour. If it had been taken to completion, it would probably have been the first passenger aircraft to have been flown completely by remote control.

The French proposed the Hermes space plane, which would have been launched vertically from atop a specially adapted booster. In the 1980s France developed the Star-H space plane. The Germans invested considerable development resources on the Sänger II, which would have consisted of a small, winged space plane launched from the back of a large, piloted hypersonic aircraft. This aircraft would have been capable of carrying 130 passengers a distance of more than 8,000 miles (12,875 km) at a cruising speed of Mach 4.5.

Meanwhile, the Japan National Space Development Agency (NASDA) had put considerable time and effort into its H2 Orbiting Plane (HOPE), a vehicle similar to the Hermes that would have been launched atop the Japanese H-II booster.

Below left: Artist's rendering of the X-43B, which may fly by the year 2010.

AVIATION IN OTHER WORLDS

Balloons would be ideal vehicles for the exploration of many of the worlds in the solar system—at least those in possession of a substantial atmosphere. Balloons have already been used to explore Venus. In 1984 the Soviet Union launched the *Vega 1* and *2* probes to Venus. These not only sent landing craft to the surface but also released French-made atmospheric balloon probes that drifted through Venus's sky while sending data back to Earth. Astronomers now have plans for a robotic blimp that will be able to navigate through the upper layers of Venus's clouds, sending back data for weeks or even months. Balloons or blimps may also be released in the atmospheres of Jupiter and Saturn.

Glossary

AERODYNAMICS. The field of science that studies how air flows and how forces act upon objects moving through air.

AERONAUTICS. The study of flight and the science of building and operating aircraft.

AILERONS. Control surfaces on the trailing edge of each wing used to make an airplane roll.

AIRFOIL. An object, such as a wing, with a special shape designed to produce lift efficiently when the object is moved through the air.

AVIATION. The development and use of aircraft. There are three types of aviation: general, commercial, and military.

BIPLANE. An airplane with two sets of wings.

CANARD. Two small projections on either side of an aircraft, mounted near the nose, that increase horizontal stability.

COMMERCIAL AVIATION. The business of operating aircraft that carry passengers or freight.

DELTA WING. A swept-back wing that looks like a triangle from above.

DIHEDRAL ANGLE. The upward angle of the wings of an aircraft formed where they connect to the fuselage.

DRAG. The force that resists the motion of aircraft through the air.

ELEVATORS. Control surfaces on the horizontal part of an aircraft's tail used to make it pitch.

ENGINE. A machine that uses combustion to create energy.

ENGINEER. Someone who designs and builds mechanical or electrical devices.

FEDERAL AVIATION ADMINISTRATION (FAA). A government agency, under the U.S. Department of Transportation, that oversees all aviation within the United States, including airport safety, air traffic control, licensing of pilots, inspection of aircraft, and the investigation of accidents.

FLAPS. Movable parts at the trailing edge of an aircraft wing that are used to increase lift at slower airspeeds.

FORWARD-SWEPT WING. A wing that is swept toward the front of the airplane.

FUSELAGE. The body of an airplane, to which the tail and wings are attached and where the passengers and cargo are located.

GENERAL AVIATION. All air activity other than scheduled airline or air-freight operations and military flights.

HORIZONTAL STABILIZER. The horizontal part of the tail that helps to increase the stability of an aircraft.

HYPERSONIC. Velocity greater than five times the speed of sound.

JET ENGINE. An engine that works by creating a high-velocity jet of air to propel an aircraft forward.

LIFT. A force, perpendicular to the airflow around an aircraft, that in normal forward flight is the force that lifts the aircraft into the air.

MILITARY AVIATION. The operation of aircraft that belong to the armed forces.

MONOPLANE. An airplane with one set of wings.

NATIONAL ADVISORY COMMITTEE FOR AERONAUTICS (NACA). A government agency started in 1917 to conduct research in aeronautics; in 1958 its name was changed to the National Aeronautics and Space Administration (NASA).

NATIONAL AERONAUTICS AND SPACE ADMINISTRATION (NASA). Agency created in 1958 to replace NACA, with the purpose of conducting research in aeronautics and astronautics.

PAYLOAD. The load carried by an aircraft, including passengers and cargo.

PITCH. A rotational motion in which an airplane turns on its lateral axis.

PROPELLER. A device consisting of airfoil-shaped blades that spin around a central hub, like a fan.

ROLL. A rotational motion in which an aircraft turns on its longitudinal axis.

RUDDER. A control surface on the trailing edge of the vertical part of the tail used to make an aircraft yaw, or turn around the vertical axis.

SPEED OF SOUND. The speed at which sound waves travel, about 770 miles per hour (1,239 km/h) at sea level.

STABILIZER. A surface that helps to provide stability for an aircraft.

STALL. A breakdown of airflow over a wing, which suddenly reduces lift.

STREAMLINE. To shape an object that creates less drag and moves smoothly and easily through the air.

SUBSONIC. Velocity less than the speed of sound.

SUPERSONIC. Velocity greater than the speed of sound.

SWEPT-BACK WING. A wing slanted toward the rear of an airplane.

THRUST. The force created by engines that pushes an aircraft through the air.

UNDERCARRIAGE. The part of an aircraft that provides support while the aircraft is on the ground.

VARIABLE-SWEEP WINGS. Wings hinged so they can be slanted forward or backward during flight.

VELOCITY. The speed at which something moves.

VERTICAL STABILIZER. The vertical part of an airplane's tail.

WIND TUNNEL. A tunnel-shaped chamber used for testing aerodynamic properties.

WING WARPING. A mechanism to provide lateral control of an aircraft through flexible wingtips.

YAW. A rotational motion in which an aircraft turns around its vertical axis.

Find Out More

BOOKS

Allen, Richard Sanders. *The Northrop Story, 1929–1939.* Atglen, PA: Schiffer Aviation History, 1993.

———. *Revolution in the Sky: The Lockheeds of Aviation's Golden Age.* Atglen, PA: Schiffer Aviation History, 1986.

Bowers, Peter M. *Curtiss Aircraft, 1907–1947.* London: Putnam, 1987.

Boyne, Walter. *The Smithsonian Book of Flight.* Washington, DC: Smithsonian Institution Press, 1987.

Brickhill, Paul. *The Dam Busters.* London: Pan, 1999.

Bryan, C.D.B. *The National Air and Space Museum.* New York: Abradale Press, 1992.

Cochran, Jackie. *Jackie Cochran: An Autobiography.* New York: Bantam Books, 1987.

Crouch, Tom. *The Bishop's Boys: A Life of Wilbur and Orville Wright.* New York: W. W. Norton, 1989.

———. *The Eagle Aloft: Two Centuries of the Balloon in America.* Washington, DC: Smithsonian Institution Press, 1983.

———. *Wings: A History of Aviation from Kites to the Space Age.* New York: W. W. Norton, 2004.

Davies, R.E.G. *Pan Am: An Airline and Its Aircraft.* Miami: Paladwr Press, 1987.

Gandt, Robert L. *China Clipper: The Age of the Great Flying Boats.* Annapolis, MD: Naval Institute Press, 1991.

Hallion, Richard P. *Test Pilots: The Frontiersmen of Flight.* Washington, DC: Smithsonian Institution Press, 1988.

Holden, Henry. *Ladybirds: The Untold Story of Women Pilots in America.* Freedom, NJ: Black Hawk Publishing, 1991.

Jenkins, Dennis R. *Space Shuttle: The History of Developing the National Space Transportation System.* Osceola, WI: Motorbooks International, 1992.

Kermode, A. C. *The Mechanics of Flight.* New York: Prentice Hall, 1996.

Langewiesche, Wolfgang. *Stick and Rudder.* New York: McGraw Hill, n.d.

Larkins, William T. *The Ford Tri-Motor, 1926–1992.* Atglen, PA: Schiffer Aviation History, 1992.

Leary, William M. *Aerial Pioneer: The U.S. Air Mail Service, 1918–1927.* Washington, DC: Smithsonian Institution Press, 1985.

Lindbergh, Charles. *The Spirit of St. Louis.* New York: Scribner, 1998.

Markham, Beryl. *West with the Night.* New York: Stewart, Tabori & Chang, 1994.

Rendell, David. *Aircraft Recognition Handbook.* London: Collins, 2002.

Rich, Doris. *Amelia Earhart: A Biography.* Washington, DC: Smithsonian Institution Press, 1990.

Rummel, Robert W. *Howard Hughes and TWA.* Washington, DC: Smithsonian Institution Press, 1991.

Saint-Exupéry, Antoine. *Airman's Odyssey.* New York: Harvest Books, 1984.

———. *Night Flight.* New York: Harvest Books, 1974.

———. *Wind, Sand and Stars.* New York: Harcourt, 1992.

Schmid, S. H., and Truman G. Weaver. *The Golden Age of Air Racing: Pre-1940.* Oshkosh, WI: EAA Aviation Foundation, 1991.

Verne, Jules. *Robur the Conqueror, or, A Trip Round the World in a Flying Machine.* Rockville, MD: Wildside Press, 2006.

Yeager, Chuck. *Chuck Yeager and the Bell X-1: Breaking the Sound Barrier.* New York: Harry N. Abrams, 2006.

WEB SITES

Air and Space magazine
www.airspacemag.com

Aviation History Online Museum
www.aviation-history.com/

Centennial of Flight
www.centennialofflight.gov

Dryden Flight Research Center
www.dfrc.nasa.gov

Experimental Aircraft Association
www.eaa.org/

First Flight Centennial Foundation
www.firstflightcentennial.org/

History of Flight
www.history-of-flight.net/index.htm

National Air and Space Museum
www.nasm.edu/

NASA
www.nasa.gov

At the Smithsonian

National Air and Space Museum

The Smithsonian Institution's National Air and Space Museum (NASM) maintains the largest collection of historic air and spacecraft in the world. It is also a vital center for research into the history, science, and technology of aviation and spaceflight, as well as planetary science and terrestrial geology and geophysics.

The collection of the National Air and Space Museum includes more than 30,000 aviation and 9,000 space artifacts. Thousands of additional artifacts—including engines, rockets, uniforms, space suits, balloons, artwork, documents, manuscripts, and photographs—document the rich history of flight.

Left: The Smithsonian Institution's National Air and Space Museum in Washington, D.C.

Below: Space Hall at the Smithsonian's National Air and Space Museum.

The National Air and Space Museum on the National Mall in Washington, D.C., has hundreds of historic artifacts displayed in 22 exhibition galleries. The collection includes such important items as the Wright 1903 *Flyer*, the *Spirit of St. Louis*, and the Apollo 11 command module *Columbia*. This facility also holds the museum's library and archives, a collection of an estimated 1.7 million photographs, 700,000 feet of motion picture film, and two million technical drawings.

The Steven F. Udvar-Hazy Center

Located in Chantilly, Virginia, the Steven F. Udvar-Hazy Center houses the thousands of aviation and space artifacts that cannot be exhibited on the National Mall. The Boeing Aviation Hangar displays aircraft on three levels. Among the aviation artifacts on display are the Lockheed SR-71 Blackbird, the Boeing B-29 Superfortress *Enola Gay*, and the De Havilland Chipmunk aerobatic airplane.

The James S. McDonnell Space Hangar opened at the center in November of 2004 and displays hundreds of spacecraft, rockets, satellites, and space-related artifacts, including the Space Shuttle *Enterprise*.

Paul E. Garber Preservation, Restoration, and Storage Facility

Located in Suitland, Maryland, the Paul E. Garber Preservation, Restoration, and Storage Facility is where the museum preserves, stores, and restores aircraft, spacecraft, and other artifacts. The facility takes up 32 buildings, some 19 of which are used to store aircraft, spacecraft, engines, and various parts.

The Steven F. Udvar-Hazy Center.

Index

Acknowledgments and Credits

The author and publisher offer thanks to those closely involved in the creation of this volume:

Tom Crouch, National Air and Space Museum, Smithsonian Institution; Ellen Nanney, Senior Brand Manager, Katie Mann, and Carolyn Gleason with Smithsonian Business Ventures; Collins Reference executive editor Donna Sanzone, editor Lisa Hacken, and editorial assistant Stephanie Meyers; Hydra Publishing president Sean Moore, publishing director Karen Prince, senior editor Molly Morrison, art director Brian MacMullen, designers Ken Crossland, Erika Lubowicki, production editors Eunho Lee and Lee Bartow, editorial director Aaron Murray, picture researcher Ben DeWalt, editors Rachael Lanicci, Suzanne Lander, Andy Lawler, Gabrielle Kappes, and Michael Smith, proofreader Glenn Novak, and indexer Jessie Shiers.

Credits

The following abbreviations are used: BS—Big Stock Photo; FI—FI; IO—Index Open; iSP—© iStockPhoto.com; JI—© 2007 Jupiterimages Corporatin; LOC—Library of Congress; NASM—National Air and Space Museum; NPS—National Park Service; PR—Photo Researchers, Inc.; SI—Smithsonian; SPL—Science Photo Library; SS—ShutterStock; Wi—Wikimedia

(t=top; b=bottom; l=left; r=right; c=center)

The World in the Air
ii US Airforce iii Wi iv NASA/Langley Research Center v NASA/Dryden vi AP Photo 1 US Air Force 3 US Air Force 4 US Air Force 5 US Air Force 6 NASA 7 NASA 8 NASA/MSFC 11 NASA/Dryden 12 NASA/Dryden 13t NASA/Dryden 13b US Dept. of Defense 14t NASA/Dryden 14b US Dept. of Defense 15 US Dept. of Defense 16 Hydra Publishing 17t Hydra Publishing 17b Hydra Publishing

Chapter 1: The First into the Air
18-19 SI 20 SI 20 LOC 22bl LOC 23tl LOC 23br Wi 24 LOC 25 LOC 26 LOC 27 LOC 28 SPL/Jean-Loup Charmet 29tc LOC 29br LOC 30 LOC 31 LOC 32 LOC 33bg LOC 33 LOC 34 LOC 35 LOC 36tl LOC 36br AP PHOTO 37 SS/Charles Shapiro 38 SI/Octave Chanute 39bg LOC 39r LOC 40 LOC 41 SS/Luisa Fernanda Gonzalez 42 SI 43 LOC 44 SI 45 SI

Chapter 2: The First Birdmen
46 inset LOC 46 LOC 48 SI 49bl LOC 49tr LOC 50 SI 51 LOC 52-53bg LOC 52 LOC 53 SI/NASM/Mark Avino 54 LOC 55 LOC 56bl LOC 56tr Rod Filan 57 Rod Filan 58 LOC/Ernest L. Jones Collection 59tl SI/NASM 59br SI/NASM 60 SI/NASM 61tl LOC 61br LOC 62 LOC 63tc SI/NASM 63br LOC 64 NASA 65tl US Air Force 65br SI/NASM 66 Wi 67bl Wi 67br Adrian Pingstone

Chapter 3: War in the Air
68 inset SI/Lt. Col. S.F. Watson/US Army 68 AP/US Navy 70 AP 71 SI/NASM 72t LOC 72b WI 73 SI/NASM 74 courtesy of Ron Miller 75 courtesy of Ron Miller 76 courtesy of Ron Miller 77 LOC 78 AP Photo 79 SS/Sergei A. Tkachenko 80tl SS/Karen Hadley 80cr Wi 81tl LOC 81bg AP Photo 82bl AP Photo 82cr Wimedia Commons 83 SS/Craig Mills 84 AP Photo 85 SI/NASM 86tl SI/NASM 86cr NASA DRYDEN 86b 87tl AP/US Navy 87c AP Photo

87br SI/ J.L. Hussey 88tl SI/NASM 88bl NASA Dryden 89tr Nasa Dryden 89b SI/NASM 90bl SI/NASM 90tr NASA 91 SS/U-96 92 SI/NASM 93 SI/NASM 94 SS/Graham Bloomfield 95 IO/John Luke 96 US Air Force 97tl US Air Force 97br US Air Force

Chapter 4: Airmail
98 inset SI 98 LOC 100 SI 101tr Alaska's Digital Archives: Edward Lewis Bartlett Papers 101bc LOC 102tl SI 102br SI 103tl SI 103tr SI 104tl LOC 104br LOC 105 Adrian Pingstone 106 SS/Ivan Cholakov 107 AP Photo

Chapter 5: The First Across
108 LOC 110bl US Navy 110br SI 111t LOC 111b SI/NASM 112tl LOC 112br LOC 113tc SI 113bc SI/NASM/Eric Long 113br LOC 114tl LOC 114bl SI/NASM 114tr LOC 115bl LOC 115tr CMG WorldWide 116tl SI/NASM 116br SI/NASM 117bl SI/NASM 117tr NASA/Jim Moran

Chapter 6: The First Around
118t SI/NASM 118b LOC 120cl SI//NASM/Eric Long 120c SI/NASM/Frank Griggs 121tr SI 121b LOC 122 LOC 123tl US Air Force 123br SI 124 AP/Doug Pizac 125tr Wi 125br SI

Chapter 7: Airlines
126–127 NASA 128 inset SI/NASM 128 IS 130 LOC/Theodor Horydczak 131 LOC 132 AP/Murray Becker 133t LOC 133b LOC 133c AP/Murray Becker 134tl SI/NASM 134br Wi/KLM Royal Dutch Airlines 135tc LOC 135cr Jerry Kuntz 135bl Wi 136 British Airways/Paul Jarvis 137t courtesy of Ron Miller 137c LOC 138 LOC 139tr SI /NASM/Hans Groenhoff 139br SI/NASM 140cl Juan Trippe 140br Juan Trippe 141tl LOC 141br The Florida Memorial Project-State Library and Archives of Florida/Pan American World Airways, Inc. 142 LOC 143 LOC 144 Adrian Pingstone 145bl SI/NASM/Dane A. Penland 145tr SI/NASM/Dane A. Penland 146 NASA/Carla Thomas 147cl LOC 147bl The Florida Memorial Project- State Library and Archives of Florida/Pan American World Airways, Inc. 148tl LOC 148br LOC 149 Wi/Marc N. Weissman 150 SS/Graham Bloomfield 151 NASA 152 Wi 153 WI

Chapter 8: Exploring the World
154 inset LOC 154 LOC 156tl SI/NASM 156br SI/NASM 157tr Wi 157br US Air Force 158bl LOC 158tr SI/NASM 159 SI/NASM 160tl LOC 160b SI/NASM 161 SI/NASM 162bl SI/NASM 162br Wi/Yosemite 163tl LOC 163br SI/NASM

Chapter 9: Flying for the Fun of It
164 inset Naomi West 164 LOC 166cl LOC 166c SI/NASM 166tr SI/NASM 167bl SI/US Air Force SI/NASM 168 SI/NASM 169 SI/NASM 170tl SI/NASM 170bl SI/NASM 171 SI/NASM 172 SI/NASM 173 US Air Force 174 SI/NASM/Taras Kiceniuk 175 SS/Graham Prentice 176 SS 177 Joseph

Chapter 10: Flying Higher
178t SI 178b Science Source 180 Centennial of Flight 181 US Navy 182 AP Photo 183 AP Photo

Chapter 11: The Quest for Speed
184 inset SI/NASM 184b NASA/Bill von Ofenheim 186tl SI/NASM 186bl SI/NASM 187bl NASA 187tr Wi 187cr

US Air Force 188tr NASA/Dryden/Tom Tschida 188br NASA/GRIN/Bill von Ofenheim 189tr NASA/Dryden/Jim Ross 189b US Dept. of Defense 190 SPL/Detlev van Ravenswaay 191 SPL/Ria Novosti 191r SI/NASM 192 NASA 193tr NASA 193br Wi 194 Wi 195cr NASA 195br NASA 196t NASA 196bl NASA/Dryden/Carla Thomas 197bl NASA/Dryden 197tr NASA/Dryden

Chapter 12: Flying into Space
198 inset AP/Reed Saxon 198 NASA 200cl NASA/GRIN 200tr NASA/MSFC 201 NASA/DRYDEN 202 NASA 203tr NASA/Dryden 203br NASA/MSFC 204 NASA/MSFC 205tr NASA 205br NASA/GRIN 206tl NASA/LaRC 206bl NASA/MSFC 207bl NASA/MSFC 207tr NASA/JPL

At the Smithsonian
211 Dane A. Penland 212tr Dane A. Penland 212bl Dane A. Penland

Cover Art
Front US Airforce Back NASA

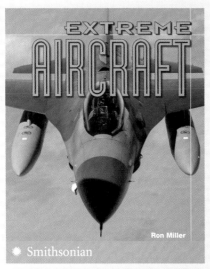
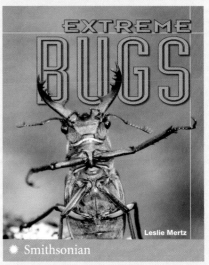
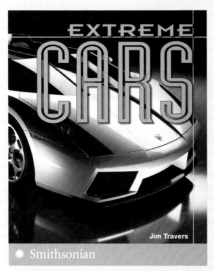
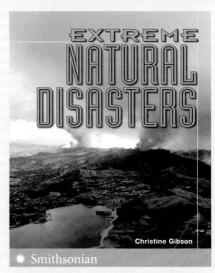

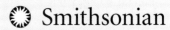